T0234318

.

# Education Outreach and Public Engagement

# Mentoring in Academia and Industry

Series Editor: J. Ellis Bell, University of Richmond, Virginia

Biology is evolving rapidly, with more and more discoveries arising from interaction with other disciplines such as chemistry, mathematics, and computer science. Undergraduate and Graduate biology education is having a hard time keeping up. To address this challenge, this bold and innovative series will assist science education programs at research universities across the country and by enriching science teaching and mentoring of both students and faculty in academia and for industry representatives. The series aims to promote the progress of scientific research and education by providing guidelines for improving academic and career building skills for a broad audience of students, teachers, mentors, researchers, industry, and more.

**Volume 1** Education Outreach and Public Engagement
by Erin L. Dolan

Erin L. Dolan

# Education Outreach and Public Engagement

 Springer

Erin L. Dolan
Department of Biochemistry
Fralin Center
Virginia Tech, Blacksburg, VA 24061
USA
edolan@vt.edu

ISBN: 978-0-387-77791-7       e-ISBN: 978-0-387-77792-4
DOI: 10.1007/978-0-387-77792-4

Library of Congress Control Number: 2008923182

Printed on acid-free paper

9 8 7 6 5 4 3 2 1

springer.com

# Preface

Whether you are at a land-grant institution with an articulated mission to translate research to the public or your child's second grade class is studying butterfly development, there are a whole host of reasons to get involved in public education about science. University expectations for promotion and tenure and recent mandates from federal agencies are requiring that scientists engage the public in a meaningful way. Yet, most scientists receive little if any preparation about how to do this. The programs, tools, and resources featured in this monograph will assist scientists and scientists-in-training in enhancing public awareness and understanding of science and considering its applications and implications. In addition, this monograph will respond to the call for outreach training put forth by Dr. Alan Lesher, CEO of the American Association for the Advancement of Science and *Science Magazine*, by addressing the notion that public engagement goes beyond enhancing the layperson's comprehension of science to engaging in real and meaningful dialogue about public "concerns and what specific scientific findings mean."

In writing this monograph, I have made a number of choices regarding the breadth and depth of the topics I have addressed. First, I have focused most of the discussion and examples on outreach and engagement with the K-12 community. Partnerships among students, teachers, and scientists are my area of interest and expertise. I was best able to draw on my experiences and those of my colleagues by concentrating on research and strategies from this arena. In addition, through conversations with informal and non-formal educators, I have learned that many of the lessons learned from K-12 outreach and engagement are applicable in their work.

Second, I intentionally chose to feature in-depth examples from individuals with whom I have personal connections, many from my own institution. At first glance, this may seem nepotistic, as there are excellent K-12 outreach and engagement activities happening across the country and around the world. Rather, my intention was to demonstrate that many grassroots efforts are borne at any single institution and your own colleagues and friends are likely to be exceptional sources of ideas and advice. In addition, many teachers and scientists are collaborating on worthy efforts to enhance science learning, yet the only way to learn about their endeavors is through word-of-mouth. As the field of K-12 outreach and engagement proliferates, it is incumbent upon all of us to present and publish our work. Thanks to the following individuals, who were willing to do so in this venue:

- Robert Choy, Exelixis, Inc.
- Patricia Caldera, Science & Health Education Partnership, University of California at San Francisco
- Daniel Capelluto, Chemistry, Virginia Tech
- Kristi DeCourcy, Fralin Center, Virginia Tech
- Carla Finkelstein, Biological Sciences, Virginia Tech
- Laura Gibson, Science & Health Education Partnership, University of California at San Francisco
- Glenda Gillaspy, Biochemistry, Virginia Tech
- Andrew Grillo-Hill, Science & Health Education Partnership, University of California at San Francisco
- Richard Helm, Biochemistry, Virginia Tech
- Charles Jervis, Auburn High School, Riner, Virginia.
- Judith Jervis, Biochemistry, Virginia Tech
- Chantell Johnson, Science & Health Education Partnership, University of California at San Francisco
- John Kowalski, Roanoke Valley Governor's School, Roanoke, Virginia.
- Cheryl Lindeman, Central Virginia Governor's School, Lynchburg, Virginia.
- Katherine Nielsen, Science & Health Education Partnership, University of California at San Francisco
- Jon Pierce-Shimomura, Gallo Center, University of California at San Francisco
- Malcolm Potts, Biochemistry, Virginia Tech
- Joseph Stepans, College of Education, University of Wyoming
- Anne Sylvester, Molecular Biology, University of Wyoming
- Dorothea Tholl, Biological Sciences, Virginia Tech
- Boris Vinatzer, Plant Pathology, Physiology & Weed Science, Virginia Tech

Finally, I would like to thank all of the teachers, scientists, students, outreach and partnership colleagues, education researchers, and others who have engaged willingly in many varied and thought-provoking discussions about science teaching and learning across the K-20+ continuum, including Carol Brandt, Julia Grady, Deborah Johnson, Susan Kirch, David Lally, Michael Lichtenstein, Christine Luketic, Nancy Moreno, Katherine Nielsen, J. Kyle Roberts, Kathryn Smith, Rebecca Smith, Louisa Stark, Kimberly Tanner, Dayna Wilhelm, Edward Wolfe, the Executive Committee of the National Association of Health & Science Education Partnerships, BIOTECH Project teachers, and PREP teachers and scientists.

# Contents

# Chapter 1
# Why?

## Identifying the Impetus for K-12 Outreach and Engagement

With enhanced public accessibility of scientific information, increased demand for a scientifically literate workforce and citizenry, stipulations from funding agencies to broaden the impact of science research, and changing reward systems at universities, scientists are looking for ways to engage the public in their work. Collaborations with K-12 students and teachers provide a systemic way to enhance students' interest in, attitudes toward, and understanding of science while contributing to the professional development of teachers and scientists alike.

## 1.1 Definitions and a Common Ground

"Broadening the impact of my research? Engaging the public? What does that mean?! Just give me some definitions and concrete examples, or I'm never going to know what I'm supposed to do!" I am regularly invited to give advice to scientists interested in broadening the impact of their research by working with K-12 students and teachers, and this is often how our conversations start. Confusion with terminology quickly metamorphoses into concern about how to go about this kind of work. Definitions are an important starting point because they allude to the philosophies and corresponding approaches to science learning involving individuals other than undergraduate and graduate students. The web-accessible Free Dictionary (http://www.thefreedictionary.com) offers the following definitions:

- *Engagement:* Something that serves to engage; a pledge.
- *Extension:* A program in a university, college, or school that offers instruction, as by television or correspondence, to persons unable to attend at the usual time or in the usual place.
- *Outreach:* A systematic attempt to provide services beyond conventional limits, as to particular segments of a community.
- *Partnership:* A relationship between individuals or groups that is characterized by mutual cooperation and responsibility for the achievement of a specified goal.
- *Service:* Offering services to the public in response to need or demand.

E.L. Dolan *Education Outreach and Public Engagement*,
DOI: 10.1007/978-0-387-77792-4_1 © Springer Science+Business Media, LLC 2008

Scholars and practitioners specializing in K-12 outreach, partnership, and public engagement would likely quibble with these less nuanced definitions. Yet, they belie the theoretical and strategic differences along a continuum of associations between K-12 and research institutions. Hewlett Packard's continuum describing university-industry partnerships serves as a model for considering K-12-university relationships (Table 1.1; Johnson 2003). At one end of the continuum is any activity that suggests awareness of and interest in reaching a community beyond "conventional limits," for example, giving a guest lecture in a local middle school, judging a science fair, or providing materials such as test tubes or chemicals. Initial forays into work with the K-12 community often include this type of brief, one-time-only experiences that may or may not be connected with other classroom activities. As individuals and institutions become more invested, they develop an awareness of each other's needs and resources. As a result, collaborative activities can be built into the day-to-day happenings of the partners. For example, a teacher may join a scientist's lab for a summer research internship, or a scientist-in-training (e.g., a graduate student or post-doctoral fellow) may visit an elementary classroom on a regular basis to co-teach science. These longer-term, integrated efforts are considered strategic partnerships, through which all partners are involved from beginning to end with the intent of mutual contribution and benefit.

The phrase "K-12 outreach and engagement" (K-12 O&E) will be used here to describe any work involving individuals in the scientific community (e.g., practicing scientists, scientists-in-training, retired scientists, etc.) and individuals within

**Table 1.1** The continuum of university-based K-12 education outreach and engagement activities

| Table 1.1 | | | | | |
|---|---|---|---|---|---|
| **Outreach Activities** | Providing materials | Guest lectures | Teacher professional development workshops | Research experience for teachers | Ongoing collaborations |
| | Providing information | Ask-a-scientist | Equipment loan  Co-teaching | Research experience for students | Teacher / scientist exchange programs |
| | Judging science fairs | Advising on / providing materials for science fairs | Mentoring students in science fairs | Support for teachers making conference presentations | Outreach training for scientists (e.g., GK-12) |
| | **Awareness** | **Involvement** | **Support** | **Sponsorship** | **Strategic Partnership** |

the K-12 education system (e.g., students, teachers, and school and district administrators). Certainly, there are mechanisms for enhancing public interest in, awareness of, and understanding about science beyond K-12 schools. For example, science centers and museums and other informal venues like Café Scientifique serve to engage the public in learning about science and discussing scientific findings for decades beyond formal schooling. In addition, other non-formal programs and settings, like 4-H, after-school science clubs, and public seminar series, use strategies for teaching science that are commonly used in classrooms. Yet, the K-12 education system is perhaps the largest and most systemic way to reach all U.S. citizens, and many of the lessons learned from K-12 O&E that are described here could be translated to better fit informal or non-formal scenarios.

K-12 O&E will be used here as a catchall for the sake of simplicity and familiarity and because they most fully represent the activities described throughout this text. "Outreach" suggests exactly that: reaching out, in this case, beyond the boundaries of the research institution. "Engagement" is included to emphasize that, in contrast to unidirectional provision of services, K-12 O&E should seriously consider the needs, interests, and offerings of the K-12 community. In addition, there are commonalities across the continuum of activities described in Table 1.1 that warrant discussing them as a whole. All require identifying the needs, interests, and resources of those involved (Chap. 2), developing infrastructure and programming that matches needs and resources (Chap. 3), and evaluating to inform the development and revision of programs and products and to document their impact (Chap. 4).

It is important not only to note what K-12 O&E is, but also what it is not. It is not service, or the day-to-day activities of being a good university citizen. While sitting on committees or performing departmental chores and other civic duties are essential activities, outreach differs by connecting "... directly to one's special field of knowledge and relat[ing] to, and flow[ing] directly out of, this professional activity" (Boyer 1990). In his text *Scholarship Reconsidered: Priorities of the Professoriate*, Boyer describes the scholarship of application and distinguishes it from good citizenship by framing it with respect to these questions:

> How can knowledge be responsibly applied to consequential problems? How can it be helpful to individuals as well as institutions? Can social problems themselves define an agenda for scholarly investigation? (*Boyer 1990*).

Most scientists are not interested in developing a new line of scholarly inquiry, but can use these questions to guide their thinking about their role in K-12 O&E in a broad way. All life scientists have content knowledge to share, not just within their specific discipline. Most have some working knowledge of physics, chemistry, and other scientific disciplines developed through undergraduate and graduate coursework or through the increasingly inter-disciplinary practice of science. In addition, scientists have habits of mind and an understanding of the nature of science (i.e., what science is) that can be shared with a broader community that may never take another science class beyond high school.

## 1.2   Motivations: Context, Carrots, and Sticks

Goals for science education span the gamut, from preparing the future scientific and science teaching workforce to ensuring a scientifically literate citizenry. The rationales for K-12 O&E are similarly diverse, from expanding students' knowledge of science concepts to mentoring students in their pursuit of science careers to modeling scientific thinking and problem solving to providing a mechanism for science teachers to stay up-to-date on current research. The overarching idea of O&E has a long and sustained history in the land-grant college system. Written in 1862, the First Morrill Act regarding land-grant colleges states that the mission of these institutions is to

> …promote the liberal and practical education of the industrial classes in the several pursuits and professions in life (U.S. Congress 1862).

This mission has been broadly interpreted to include not only the offering of practical educational opportunities to the general public, but also to apply the learning developed within higher education to broader public benefit. The land-grant movement has since expanded to include historically black colleges and universities and Tribal Colleges, and its curriculum has metamorphosed from traditional farming, mechanics, and military arts to experimental agriculture, engineering, and other applied sciences. The impetus for these changes has been the needs and interests of the public. Some argue that the benefits of O&E that are realized by universities in general and scientists in particular are sufficiently compelling. For example, Schoenfeld describes how the "Wisconsin Idea" is embodied in the University of Wisconsin extension system by equating of the boundaries of the university with the boundaries of the state:

> university outreach is an institutional state of mind which views the university not as a place but as an instrument. In actual operation, outreach leaders… seek to identify public problems, to stimulate public awareness and concern, to interpret public educational needs to the university, to focus university skills and resources upon them, and thence to translate university insights into a wide range of formal and informal educational service activities throughout a state or region. The outreach mission, in essence, is to bring campus and community into fruitful juxtaposition, thereby immeasurably enriching the life of both (Schoenfeld 1977).

Others point out that it is the responsibility of scientists, especially those whose work is supported by public funds, to share their findings with their fellow citizens in a way that makes the benefits concrete (e.g., Market and Opinion Research International 2000). Regardless of the underlying rationale, how this mission manifests in the day-to-day activities of the university and its faculty varies with the needs, resources, and interests of the individuals and institutions involved.

In an effort to foster scientists' participation in K-12 O&E, universities, funding agencies, and other policymakers have developed relevant expectations and rewards. While some institutions encourage employees to engage the public by providing incentives, others hire personnel with an explicit expectation related to O&E programming and scholarship. As a result, an expanding number of academic institutions are putting outreach-related metrics in place to involve scientists in working with K-12 schools, rewarding these activities through promotion and tenure.

For example, Promotion and Tenure Guidelines within the University of Arizona's College of Agriculture and Life Sciences stipulate consideration of outreach activities alongside teaching and research:

Outreach is a form of scholarship that cuts across teaching and research/creative activity. It involves delivering, applying, and preserving knowledge for the direct benefit of external audiences in ways that are consistent with University and unit missions. The application of one's expertise to issues in the community is encouraged and often generates research ideas and contributions (University of Arizona Office of the Provost 2004).

If the individual's job involves outreach, then presentations, publications and other output may be evaluated with their other teaching and research activities. For these activities, the candidate should be developing a reputation of excellence among peers (University of Arizona College of Agriculture and Life Sciences 2001).

The University of Arizona has also established a Science Education Promotion and Tenure Committee (SEPTC; University of Arizona 1997) to assist science departments in evaluating faculty whose primary appointment involves the preparation and professional development of science teachers.

A number of institutions offer promotion and tenure for outreach scholarship. For example, University of Wisconsin faculty must demonstrate excellence in a primary area (i.e., research, teaching, or outreach/extension) and significant accomplishment in one of the two remaining areas as grounds for promotion or tenure. Faculty with at least a 50% outreach/extension appointment may use achievement in outreach per se as this basis (University of Wisconsin-Madison Faculty Division of the Biological Sciences 2007). In addition, individuals who have a primary appointment in research can be rewarded for their outreach work as their complementary area of significant accomplishment. Virginia Tech's promotion and tenure policy also offers guidelines for acknowledging a faculty member's accomplishments in research, instruction, and outreach in accordance with his or her assignment (Virginia Tech Office of the Provost 2007). These changes may not seem revolutionary, especially among those familiar with promotion and tenure in the extension system (Olsen 2005). Yet, other university departments are embracing these reward systems, appointing faculty whose primary responsibility is outreach and evaluating them accordingly, rather than with respect to metrics more appropriate for faculty engaged in bench, field, or theoretical research.

Other organizations are dedicating significant resources to building K-12 O&E capacity of current and future science faculty. Many of these efforts have been initiated in response to expectations posed by extramural funding agencies. For example, the National Science Foundation (NSF) will not consider any proposal for funding that fails to address explicitly how the investigators will broaden the impact of their research through education, outreach, or mentorship. Bob Eisenstein, Chair of the Panel on Public Affairs of the American Physical Society, was at NSF when the criterion was first put in place in the mid-1990s. He explained that the criterion is meant to serve two purposes: first, it forces scientists to think more carefully about the ways in which their work impacts society, and second, it helps provide the public with more information about what scientists are doing (American Physical Society 2007). An obvious and widely available venue for reaching the public is the

K-12 school system, as illustrated by the repeated inclusion of K-12 audiences in example activities (NSF 2007).

Foundations and other charitable organizations have also called for colleges, universities, businesses, and other community groups to play substantive roles in K-12 science education. For example, the Bill and Melinda Gates Foundation recommends involvement in after-school programming, tutoring, and summer employment as well as in providing space, materials, and internships (Bill and Melinda Gates Foundation 2007). These types of assistance are intended to accomplish two complementary goals: encouraging students to work hard and stay in school and reforming our nation's system of education to make it more compelling, relevant, and rigorous.

Other benefactors provide financial incentives specifically for the involvement of scientists in K-12 schools. For example, the Wellcome Trust has established *Engaging Science*, a grants program designed to support national and international efforts to engage the public in biomedical science, as well as better understand how this is accomplished (Wellcome Trust 2007). In addition, the Howard Hughes Medical Institute (HHMI) Professors program supports efforts to reform undergraduate teaching and learning for students majoring in science and other disciplines (HHMI 2007). Many of the individuals in HHMI's Society of Professors also engage pre-college audiences through their undergraduate work or through complementary efforts. Thus, the "carrots" available to scientists interested in outreach are multiplying.

## 1.3   Motivations: Personal and Public Gains

The availability of concrete rewards like tenure, promotion, and extramural funding certainly encourages interested scientists to get involved in K-12 education. High profile researchers, science and education policymakers, and legislators have also made pleas in support of scientists' engagement with the public, with aims to encourage science-related public dialogue, share new findings, recruit the next generation of scientists, ensure scientific literacy of U.S. citizenry, or respond to public needs via the design and conduct of research (Alberts 2005; Leshner 2007; NRC 1996; Shalala 1991; Wheeler 1998). Most of these calls for action are based on concerns about the "state of the system," for example, statistics regarding students' dismal test scores, students' failure to pursue educational and professional opportunities in science, teachers' lack of preparedness to science, or even the dearth of qualified applicants for positions in science industry and teaching. Although these numbers and their origins and implications are worth considering, what is less clear is whether or how scientists' involvement in K-12 O&E will remedy the situation. Excitingly, the results of research and evaluation of university-based K-12 O&E efforts are beginning to reveal the benefits that may spawn greater involvement (Table 1.2).

Some scientists become interested in K-12 education when their own children are in school, with an aim to enrich their own children's education. Indeed,

**Table 1.2** Benefits of K-12 outreach and engagement activities

Students
- Improved science achievement
- Positive attitudes towards science
- Increased interest in science and scientific careers
- Access to science role models
- Enhanced attentiveness and involvement in science class
- Better science class attendance

Teachers
- Enhanced science learning for their students
- Reduced classroom management issues
- More in-depth understanding of science content and process
- Increased confidence about own science knowledge
- Increased enthusiasm about science
- Greater likelihood to use inquiry-oriented instructional strategies
- Access to a larger and more diverse professional network
- Access to science role models

Scientists
- Improved teaching and communication skills
- Improved public speaking skills
- Enhanced reflection on own teaching practice
- Access to teaching role models
- Better awareness of the interests, abilities, and needs of students entering undergraduate programs
- Better understanding of own research
- Insights into research on teaching and learning

Scientific Community
- Recruitment and preparation of future scientists
- Development of scientifically literate citizenry
- Expansion of public support for science
- Reduction of negative stereotypes about science and scientists

scientists can influence students' achievement in science and math and instill positive attitudes toward science and scientists by participating in classroom activities (Laursen et al. 2007; Spillane 2004; Waksman 1999). A new voice in the classroom can enhance students' attentiveness and interest in science class, reducing classroom management issues and prompting students to be more involved in class activities (Laursen et al. 2007; Spillane 2004). Other benefits have been proposed that have yet to be published in the literature, most likely because they are more difficult to capture. For example, scientists are in a position to provide up-to-date scientific information that learners might find useful in decision-making. Similarly, scientists can serve as role models, not only for pursuing post-secondary education or scientific careers (Bruce et al. 1997; Speizer 1982), but also for how to reason scientifically, develop explanations based on evidence, and demonstrate a comfort with ambiguity that is an essential element of doing science.

Teachers are often motivated to collaborate with scientists because of the benefits realized by their students, rather than by any advantages they experience directly. Yet, teachers can gain a more in-depth understanding of science content and process through their work with scientists (Lockhart and Le Doux 2005; Morrison and Estes 2007). Notably, teachers' participation in science research yields positive outcomes for students and teachers alike. For example, one multi-program study demonstrated that students of teachers who had participated in scientific work experiences, usually summer internships in research labs, had more positive attitudes about science, better science class attendance, and higher science achievement test scores when compared to students taught by a similar group of teachers who had not participated in research (i.e., teachers who have similar amounts of training and who teach similar students in similar classroom, school, and community contexts; Dubner et al. 2001; S. Silverstein, J. Dubner, pers. comm. SWEPT multi-site student outcomes study; http://www.swept-study.org, accessed 12/3/2007). Although the elements of teachers' research experiences that may lead to these outcomes have not been identified, results from a series of pre/post case studies are consistent with the interpretation that teachers become more confident about their science knowledge, more enthusiastic about science, and more apt to use inquiry instruction in their classrooms following a research experience (Blanchard 2006; Dutrow 2005; Strand et al. 2005).

Teaching as a profession is isolating, and science research internships are a mechanism for teachers to become part of a group (Strand et al. 2005). The immediate professional network of a research group presents opportunities to collaborate with a variety of scientists who have different perspectives, for example, professors with teaching responsibilities or undergraduates with recent experience making the transition from high school to college. Because teachers rarely get to observe others' teaching or research or discuss strategies with co-workers, scientific work experiences provide a ready-made network for collegial discussion. In some venues, this network extends beyond the immediate research group to other groups that are in the same department or building or who are regular collaborators within and across disciplines. Teachers can tap this extended network for expertise and access to materials, supplies, and ideas for classroom-based science learning. By collaborating with teachers in scientific work, scientists can bring the professional status afforded to individuals with technical experience to issues in K-12 education. Teachers can gain some of this status themselves through the development of technical expertise. Finally, as scientists expand their awareness and understanding of the opportunities and challenges in pre-college education, they can be better advocates for K-12 science learning (e.g., Bower 1996; McKeown 2003; Schultz 1996).

Benefits to scientists are both personal and professional. Many scientists get involved in K-12 O&E because it is personally rewarding to see students building enthusiasm about science or to witness students having a "eureka!" moment in science learning. A number of studies have documented the professional benefits scientists have realized through their outreach work. First, scientists can develop and fine-tune their teaching skills through K-12 O&E (Busch and Tanner 2006; Laursen et al. 2007). Through collaborations with K-12 teachers, scientists can discuss teaching practices and problems with a shared goal of enhancing student learning. In particular, scientists working with K-12 teachers have reported that "collaborative reflection" is particularly fruitful for informing changes in their own teaching practice (Bower 1996; Busch and Tanner 2006). By teaching in pairs or small groups, collaborators can observe how someone else would teach a concept or skill and how students may or may not be learning.

> The benefits I gained from starting to examine my own science teaching were immediate and numerous. I changed my undergraduate laboratory from a 'cookbook' exercise, designed to prove to students that what I said in class was right, to an open inquiry into neuroscience (James Bower, University of Texas Health Sciences Center at San Antonio, former co-director of the Caltech Pre-college Science Initiative (CAPSI)).

Skill development extends beyond teaching per se, to include enhanced comfort with and ability to speak publicly. Scientists are required to communicate regularly in the practice of science, initially by sharing results and experimental problems during group meetings and eventually by communicating findings to the broader scientific community, both orally and in writing. Outreach activities provide scientists with a venue for practicing scientific communication and with experts, in the form of science teachers, who think about and practice communicating science to non-technical audiences (Busch and Tanner 2006).

The utility of teaching and communications skills extends to professional pathways other than academia. A variety of other jobs in science require strong teaching and communication skills, for example, marketing and sales in biotechnology or pharmaceutical companies, public relations in scientific societies, and writing or editorial positions with scientific journals. For individuals earlier in their scientific training (e.g., graduate students or post-doctoral fellows), outreach can provide a good venue to develop and fine-tune communication skills useful for landing positions that require them.

Scientists' interest in their own research can be rekindled through K-12 O&E. Children and their teachers bring creative, original, and big picture ways of thinking about science as they are not steeped in it on a daily basis and they have not yet narrowed their interests or expertise. Youthful exuberance and enthusiasm about using the tools and materials of science can be contagious (Tanner 2000; Busch and Tanner 2006). Scientists can use outreach experience as venues for reflection about the needs, interests, and preparedness of undergraduate students that can inform university-level instruction (Busch and Tanner 2006; Morrison and Estes 2007). Finally, some scientists even come to understand their own science better through the processes of preparing to teach and teaching it.

O&E activities, especially within a larger program or infrastructure that encourages reflection about teaching, also provide opportunities for scientists to benefit from what is known from theoretical and practical research in K-12 education. For example, K-12 O&E provides a venue for understanding the diversity of students in K-12 schools as well as how to engage students with diverse learning styles (e.g., Cassidy 2004; Coffield et al. 2004). Other studies examine assessment in K-12 schools, including its power and pitfalls. Instructors can use results from student assessments to make decisions about what to teach next (i.e., formative assessment), measure of the effectiveness of a lesson or pedagogical strategy, or reveal students' misconceptions that should be explicitly addressed in subsequent learning experiences (e.g., Dochy and McDowell 1997; Wolf et al. 1991).

More generally, the scientific community aims to benefit from K-12 O&E, through the recruitment and preparation the next generation of scientists, development of a scientifically literate citizenry, and mitigation of negative stereotypes about science and scientists. First, scientists can serve as role models for pursuing educationand careers in science (Holton 1992). Notably, Tai and colleagues (2006) determined that eighth graders who expected to have a career in science by age 30 were 1.9 times more likely to complete a baccalaureate degree in life science and 3.4 times more likely to complete a baccalaureate degree in physical science or engineering than those who did not anticipate entering a science career. Although this does not provide evidence of a causal relationship between the presence of science role models and eventual pursuit of a scientific career, it does suggest the value of affecting students' interests in science careers early in their education.

Although some K-12-university partnerships are designed to prepare students for further education in undergraduate settings, the primary goal of U.S. science education is to prepare a citizenry who can consider scientific information and make science-related decisions in an informed way (AAAS 1989; NRC 1996). Approximately half of U.S. citizens do not pursue college education, and 75% do not complete a college degree. Just as universities are in the business of preparing graduates for future employment in a variety of career paths, K-12 schools are in the business of teaching all pre-college students, including those who do not pursue undergraduate or graduate educations. It is impossible to teach any learner all of the science concepts and processes they will ever need to know to understand and critically evaluate the issues that arise as a result of new developments in agriculture, medicine, and environmental science. Rather, success will be measured not in knowledge acquired now, but in the development of students' reasoning abilities and the encouragement of lifelong dialogue about science and its applications and implications.

Finally, although incidents of scientific misconduct are few and far between, they garner much media attention when they do occur. Positive, ongoing relationships among scientists and the general public can help mitigate some of the suspicion or ill feelings spawned by isolated cases of misconduct.

## 1.4 Challenges

Research and evaluation of K-12 O&E efforts are also revealing the difficulties and dilemmas that individuals and institutions are facing in the process. Teachers and scientists alike face real and perceived obstacles to successful collaboration, with each other and in their own professional communities. For example, end-of-course testing encouraged by the No Child Left Behind Act (2001) has strengthened teachers' concerns about "covering" course content and their views about the need to opt for efficient didactic teaching methods and refrain from incorporating scientific inquiry in their courses. Teachers' primary concern is that the students will not have sufficient time to learn the required course content if they are engaged in inquiry (Hofstein and Lunetta 2004; Magnusson and Palincsar 2005; Tobin and McRobbie 1996; Wallace and Kang 2004). Some teachers feel pressure from administrators, students, parents, university faculty, and even other teachers to ensure their instructional strategies maximize content "coverage" (Brickhouse and Bodner 1992; Crawford 1999; Duschl 1988; Fensham 1993; Marx et al. 1994; NRC 2000; Tobin and McRobbie 1996; Wallace and Kang 2004). Scientists who are willing to help teachers face these challenges may find their involvement more welcomed.

Scientists also face issues in their communities. Some decide not to get involved because it will interfere with their research or are discouraged by colleagues for choosing to spend time on K-12 O&E rather than research (Andrews et al. 2005; Laursen et al. 2007). On the contrary, scientists-in-training who participate in K-12 O&E find ways to be more organized or efficient with their time. For example, graduate students who invest 10+ h/week in K-12 O&E through their participation in an NSF-funded Graduate Teaching Fellows in K-12 Education program (GK-12; http://www.nsf.gov/funding/pgm_summ.jsp?pims_id=5472&org=DGE) report that their research does not suffer (Spillane 2004). Other graduate students and post-doctoral fellows have found that the time they spend on research is actually more productive (Tanner 2000). Yet, most faculty are rewarded based on the publication of research papers reviewed by their peers and the quantity of grant dollars awarded by federal programs advised by experts in the scientific community (Calleson et al. 2002; Calleson et al. 2005; Dodds et al. 2003). Given the hard work and time necessary for successful K-12 O&E, activities ultimately must be framed by an individual's priorities as well as those of their institutions (Scowcroft and Knowlton 2005), balanced by the needs and interest of K-12 students, teachers, and schools.

The challenge…is how to identify problems, define goals, and design solutions that are relevant and informative to both practitioners and scholars (Sandmann et al. 2000).

Some would argue that there is no way to align scholarly activities valued by peers and those valued by the public. Perhaps outreach can serve as a context for scholars to *listen* to public concerns and interests and *discuss* with the public the justification for considering certain directions. This point is emphasized by Alan Leshner, CEO of the American Association for the Advancement of Science (AAAS) and executive publisher of *Science*:

…scientists must engage more fully with the public about scientific issues and concerns that society has about them… the notion of public engagement goes beyond public education. We must have a genuine dialogue with our fellow citizens about how we can approach their concerns and what specific scientific findings mean (Leshner 2007).

One of the biggest challenges in K-12 O&E involves finding ways to communicate. Understanding one another's language can be daunting prospect, fraught with stumbling blocks. For the most part, teachers and scientists alike are unfamiliar with each other's discipline, including its values, practices, and jargon. Thus, both can fall into the trap of failing to recognize their own jargon or assumptions. Life scientists are familiar with the old adage that learning biology, which is replete with discipline-specific terminology, resembles learning a foreign language. Education has its own

**Table 1.3**  Resources for responding to the challenges of K-12 outreach and engagement

| Challenge | Advice and strategies |
| --- | --- |
| Maintaining alignment with science education standards | Take a new view of the standards; consider other venues for science learning (Laursen 2006) |
| | Remember that understanding the processes and nature is a standard (Laursen 2006; Munn et al. 1999) |
| Navigating the differences in the cultures, language, and communication styles of teachers and scientists | Focus on mutual learning; acknowledge distinctions and commonalities (Bower 1996; Dolan and Tanner 2005; Schultz 1996; Tanner et al. 2003) |
| | Emphasize the collaborative nature of relationships between teachers and scientists (Bottomley et al. 2001; Moreno 2005) |
| Finding time to do the work | Make the commitment finite and do-able (Lally et al. 2007) |
| | Make every effort to reduce extra work required of teachers, including providing necessary materials in a timely fashion (McKeown 2003; Moreno 2005; Elgin et al. 2005) |
| | Consider the variety of roles scientists can play (Bybee and Morrow 1998; Morrow 2000; Munn et al. 1999) |
| | Articulate how the proposed activities contribute to the professional development of teachers and scientists (Tanner and Allen 2006) |
| | Save time and avoid "reinventing the wheel" by joining an existing program (Dolan et al. 2004; Dolan and Tanner 2005; Munn et al. 1999) |
| Fitting into a crowded curriculum | Articulate how the proposed activities help meet the standards (Lally et al. 2007; McKeown 2003; Moreno 2005) |
| Ensuring buy-in | Initiate discussions with teachers, administrators, and other stakeholders early on to ensure that needs are met and key stakeholders are onboard (McKeown 2003; Moreno 2005; Tomanek 2005) |
| | Involve only those who want to participate (Moreno 2005) |

language, and teachers' and scientists' conversations about teaching and learning can be facilitated through the use of common vocabulary (Dolan 2007; NAS 1996). A number of tools have been developed to assist non-experts in learning the jargon, including online and hard copy dictionaries and glossaries. *The Lingo of Learning: 88 Terms Every Science Teacher Should Know* (Colburn 2003) and *EdSpeak: A Glossary of Education Terms, Phrases, Buzzwords, and Jargon* (Ravitch 2007) define education practice terms and research vocabulary most relevant to schools and classrooms. *EdSpeak* also includes a handy list of acronyms for phrases, policymaking groups, funding agencies, accrediting bodies, and other relevant organizations.

Similarly, teachers can become frustrated with the sheer volume of science vocabulary and the rapidity with which is it used (Morrison and Estes 2007). Just as most scientists are novices to K-12 science education, most K-12 students and teachers are comparative novices in science. Research on expertise yields insights into the challenges faced by novices, or individuals who are new to a discipline (Bransford et al. 1999). Specifically, novices have yet to develop the schemas that help experts organize their knowledge, and thus are less facile at recognizing salient information or inferring meaning from context. Both communities would be well served by reflecting about what is truly important to know and how to discuss it as inclusively as possible. Those who are more expert can play a critical role in helping novices identify and filter out peripheral ideas and information. Project 2061 of the American Association for the Advancement of Science (http://www.project2061. org/) has published *Designs for Science Literacy* (2001), a valuable tool for accomplishing this in K-12 science curriculum planning.

Excitingly, the expansion of K-12 O&E efforts means that tools are being developed to foster their initiation, encourage their sustainability, and help those involved to overcome barriers (Table 1.3). For example, the National Academy of Sciences initiated RISE, or Resources for Involving Scientists in Education (NAS 1996), which has a web-based archive of ideas, strategies, and reflective essays for interested scientists. Individuals involved in K-12 O&E have also authored essays and commentaries to share lessons learned and provide advice for interested scientists (Table 1.3). In addition, new and existing activities are more often contexts for research to identify whether and how these efforts are successful, as little research to date has investigated the involvement of scientists in K-12 education directly (Tanner et al. 2003; Laursen et al. 2007).

# References

Alberts B (2005) A wakeup call for science faculty. Cell 123:739–741.

American Association for the Advancement of Science (1989) Project 2061: Science for all Americans. Oxford University Press, New York.

American Association for the Advancement of Science (2001) Project 2061: Designs for science literacy. Oxford University Press, New York.

American Physical Society (2007) NSF's "Broader Impacts" Criterion Gets Mixed Reviews. APS News 16 (6). http://www.aps.org/publications/apsnews/200706/nsf.cfm. Accessed 14 November 2007.

Andrews E, Hanley D, Hovermill J, Weaver A, Melton G (2005) Scientists and public outreach: Participation, motivations, and impediments. J Geosci Educ 53:281–293.

Bill and Melinda Gates Foundation (2007) High Schools for the New Millennium: Imagine the possibilities. http://www.gatesfoundation.org/UnitedStates/Education. Accessed 14 November 2007.

Blanchard MR (2006) Assimilation or transformation? An analysis of change in ten secondary science teachers following an inquiry-based research experience for teachers (dissertation). Florida State University, Tallahassee, Florida.

Bottomley LJ, Parry EA, Brigade S, Coley L, Deam L, Goodson E, Kidwell J, Linck J, Robinson B (2001) Lessons learned from the implementation of a GK-12 grant outreach program. Proc Amer Soc Eng Educ Ann Conf Expo Sess 1692.

Bower JM (1996) Scientists and science education reform: Myths, methods, and madness. National Academy of Science Resources for Involving Scientists in Education. http://www.nas.edu/rise/backg2a.htm. Accessed 11 May 2007.

Boyer EL (1990) Scholarship Reconsidered: Priorities of the Professoriate. Carnegie Foundation for the Advancement of Teaching, Princeton, New Jersey.

Bransford JD, Brown AL, Cocking RR (Eds.) (1999) How People Learn: Brain, Mind, Experience, and School. National Academy Press, Washington DC.

Brickhouse N, Bodner GM (1992) The beginning science teacher: Classroom narratives of convictions and constraints. J Res Sci Teach 29:471–485.

Bruce B, Bruce S, Conrad R, Huang H (1997) University science students as curriculum planners, teachers and role models in elementary school classrooms. J Res Sci Teach 34:69–88.

Busch A, Tanner KD (2006) Developing scientist educators: Analysis of integrating K-12 pedagogy and partnership experiences into graduate science training. Natl Assoc Res Sci Teach Ann Conf (San Francisco, CA, April 3–6, 2006).

Bybee R, Morrow C (1998) Improving science education: The role of scientists. Newsl Forum Educ Amer Phys Soc. http://www.spacescience.org/Education/ResourcesForScientists/Workshops/Four-Day/Resources/Articles/1.html. Accessed 3 December 2007.

Calleson DC, Seifer SD, Maurana C (2002) Forces affecting the community involvement of AHCs: Perspectives of institutional and faculty leaders. Acad Med 77:72–81.

Calleson DC, Jordan C, Seifer SD (2005) Community-engaged scholarship: Is faculty work in communities a true academic enterprise? Acad Med 80:317–321.

Cassidy S (2004) Learning styles: An overview of theories, models, and measures. Educ Psych 24:419–444.

Coffield F, Moseley D, Hall E, Ecclestone K (2004) Learning styles and pedagogy in post-16 learning: A systematic and critical review (Report No. 041543). Learning and Skills Research Centre, London.

Colburn A (2003) The lingo of learning: 88 terms every science teacher should know. NSTA Press, Arlington VA.

Crawford BA (1999) Is it realistic to expect a preservice teacher to create an inquiry-based classroom? J Sci Teach Educ 10:175–194.

Dochy F, McDowell L (1997) Assessments as a tool for learning. Stud Educ Eval 23:279–298..

Dodds J, Calleson D, Eng G, Margolis L, Moore K (2003) Structure and culture of schools of public health to support academic public health practice. J Public Health Manag Pract 9:504–512.

Dolan EL (2007) Grappling with the literature of education research and practice. CBE Life Sci Educ 6:289–296.

Dolan EL, Tanner KD (2005) Moving from outreach to partnership: Striving for articulation and reform across the K-20+ science education continuum. Cell Biol Educ 4:35–37.

Dolan EL, Soots BE, Lemaux PG, Rhee SY, Reiser L (2004) Strategies for avoiding reinventing the precollege education and outreach wheel. Genet 166:1601–1609.

Dubner J, Silverstein SC, Carey N, Frechtling J, Busch-Johnsen T, Han J, Ordway G, Hutchison N, Lanza J, Winter J, Miller J, Ohme P, Rayford J, Sloane Weisbaum K, Storm K, Zounar E (2001) Evaluating science research experience for teachers programs and their effects on student interest and academic performance: A preliminary report of an ongoing collaborative study by eight programs. Mater Res Soc Symp Proc 684E:GG3.6.1– GG3.6.12.

Duschl RA (1988) Abandoning the scientific legacy of science education. Sci Educ 72:51–62.

Dutrow JM (2005) An assessment of teachers' experiences in scientific research as a method for conceptual development of pedagogical content knowledge for inquiry (dissertation). Florida State University, Tallahassee, Florida.

Elgin SCR, Flowers S, May V (2005) Modern genetics for all students: An example of a high school/university partnership. Cell Biol Educ 4:32–34.

Fensham PJ (1993) Academic influence on school science curricula. J Curric Stud 25:53–64.

Hofstein A, Lunetta VN (2004) The laboratory in science education: Foundations for the twenty-first century. Sci Educ 88:28–54.

Holton G (1992) How to think about the 'anti-science' phenomenon. Public Underst Sci 1:103–128.

Howard Hughes Medical Institute (2007) Teaching and Research. http://www.hhmi.org/research/professors. Accessed 15 June 2007.

Johnson WC (2003) University relations: The HP model. Ind High Educ 17:391–395.

Lally D, Brooks E, Tax FE, Dolan EL (2007) Sowing the seeds of dialogue: Public engagement through plant science. Plant Cell 19:2311–2319.

Laursen SL (2006) Getting unstuck: Strategies for escaping the science standards straightjacket. Astron Educ Rev 5:162–177.

Laursen S, Liston C, Thiry H, Graf J (2007) What good is a scientist in the classroom? Participant outcomes and program design features for a short-duration science outreach intervention in K–12 classrooms. CBE Life Sci Educ 6:49–64.

Leshner AI (2007) Outreach training needed. Science 315:161.

Lockhart A, Le Doux J (2005) A partnership for problem-based learning. Sci Teach 72:29–33.

Magnusson SJ, Palincsar AS (2005) Teaching to promote the development of scientific knowledge and reasoning about light at the elementary school level. In: National Research Council, Bransford JD, Donovan MS (Eds.) How students learn: Science in the classroom. National Academy Press, Washington DC.

Market and Opinion Research International (2000) The role of scientists in public debate. Wellcome Trust: London. http://www.wellcome.ac.uk/doc_wtd003429.html. Accessed 31 October 2007.

Marx R, Blumenfeld P, Krajcik J, Blunk M, Crawford B, Kelly B, Meyer KM (1994) Enacting project-based science: Experiences of four middle grade teachers. Elem School J 94:517–538.

McKeown R (2003) Working with K-12 schools: Insights for scientists. BioSci 53:870–875.

Moreno N (2005) Science education partnerships: Being realistic about meeting expectations. Cell Biol Educ 4:30–32.

Morrison JA, Estes JC (2007) Using scientists and real-world scenarios in professional development for middle school science teachers. J Sci Teach Educ 18:165–184.

Munn M, Skinner PO, Conn L, Horma HG, Gregory P (1999) The involvement of genome researchers in high school science education. Genome Res 9:597–607.

National Academy of Sciences (1996) Resources for Involving Scientists in Education. http://www.nas.edu/rise/. Accessed 11 May 2007.

National Research Council (1996) National science education standards. National Academy Press, Washington DC.

National Research Council (2000) Inquiry and the national science education standards. National Academy Press, Washington DC.

National Science Foundation (2007) Merit Review Broader Impacts Criterion: Representative Activities. www.nsf.gov/pubs/gpg/broaderimpacts.pdf. Accessed 30 November 2007.

Olsen S (2005) County agents and university tenure and promotion systems. J Ext 43 (3). http://www.joe.org/joe/2005june/rb5.shtml. Accessed 14 November 2007.

Ravitch D (2007) EdSpeak: A glossary of education terms, phrases, buzzwords, and jargon. Association of Supervision and Curriculum Development, Alexandria VA.

Sandmann LR, Foster-Fishman PG, Lloyd J, Rahue W, Rosaen C (2000) Managing critical tensions: How to strengthen the scholarship component of outreach. Change 32:44–52.

Schoenfeld C (1977) The Outreach University. A Case History in the Public Relationships of Higher Education, University of Wisconsin Extension, 1885–1975. Office of Inter-College Programs, University of Wisconsin, Madison.

Schultz T (1996) Science education through the eyes of a physicist. National Academy of Science Resources for Involving Scientists in Education. http://www.nas.edu/rise/backg2d.htm. Accessed 11 May 2007.

Scowcroft GA, Knowlton C (2005) Proceedings of the conference on teacher research experiences (University of Rhode Island, April 24–27, 2005).

Shalala DE (1991) New paradigms: the research university in society. Teach Coll Rec 92:528–540.

Speizer JJ (1982) Role models, mentors, and sponsors: The elusive concepts. J Women Cult Soc 6:696–712.

Spillane SA (2004) Sharing strengths: Educational partnerships that make a difference. Ann Meet Amer Educ Res Assoc (San Diego, CA, April 12–16, 2004).

Strand MA, Wignall S, Leslie-Pelecky DL (2005) Research experiences for teachers in materials science: A case study. Mater Res Soc Symp Proc 861E:PP3.4.1– PP3.4.6.

Tai RH, Liu CQ, Maltese AV, Fan X (2006) Planning early for careers in science. Science 312:1143–1144.

Tanner KD (2000) Evaluation of scientist-teacher partnerships: Benefits to scientist participants. Natl Assoc Res Sci Teach Ann Conf (New Orleans, LA, April 30-May 3, 2000).

Tanner KD, Allen D (2006) Approaches to biology teaching and learning: On integrating pedagogical training into the graduate experiences of future science faculty. CBE Life Sci Educ 5:1–6.

Tanner KD, Chatman L, Allen D (2003) Approaches to biology teaching and learning: Science teaching and learning across the school-university divide: Cultivating conversations through scientist-teacher partnerships. Cell Biol Educ 2:195–201.

Tobin K, McRobbie CJ (1996) Cultural myths as constraints to the enacted science curriculum. Sci Educ 80:223–241.

Tomanek D (2005) Building successful partnerships between K–12 and universities. Cell Biol Educ 4:28–29.

United States Congress (1862) First Morrill Act. http://www.csrees.usda.gov/about/offices/legis/morrill.html. Accessed 2 December 2007.

University of Arizona (1997) SEPTC: Science and Math Education Promotion and Tenure Committee. http://www.biology.arizona.edu/raire/septc.html. Accessed 14 November 2007.

University of Arizona College of Agriculture and Life Sciences (2001) Guidelines and Criteria for Promotion and Tenure. http://ag.arizona.edu/dean/p&tguidelines.html. Accessed 9 November 2007.

University of Arizona Office of the Provost (2004) Faculty Senate-approved Recommendations: Criteria and Evaluation for Promotion and Tenure. http://academicaffairs.arizona.edu/p&t/senategd.htm. Accessed 9 November 2007.

University of Wisconsin-Madison Faculty Division of the Biological Sciences (2007) Guidelines for Recommendations for Promotion or Appointment to Tenure Rank. http://www.secfac.wisc.edu/divcomm/biological/TenureGuidelines.pdf. Accessed 14 November 2007.

Virginia Tech Office of the Provost (2007) Promotion and Tenure Guidelines 2007–2008. http://www.provost.vt.edu/tenure.php. Accessed 14 November 2007.

Waksman BH (1999) Discovery-based disclosure. Amer Sci 87:104–107.

Wallace CS, Kang N (2004) An investigation of experienced secondary science teachers' beliefs about inquiry: An examination of competing belief sets. J Res Sci Teach 41:936–960.

Wellcome Trust (2007) Public Engagement: Engaging Science. http://www.wellcome.ac.uk/funding/publicengagement. Accessed 15 June 2007.

Wheeler G (1998) The wake-up call we dare not ignore. Science 279:1611.

Wolf D, Bixby J, Glenn J, Gardner H (1991) To use their minds well: Investigating new forms of student assessment. Rev Res Educ 17:31–74.

# Chapter 2
# Who and What?

# Defining the Scope of K-12 Outreach and Engagement

Outreach to whom, where, and how? The answers to these questions are shaped initially by one's interests, experience, and expertise. Once the options are narrowed, the particular individuals, venue, and activities involved will further delineate one's K-12 outreach and engagement efforts.

## 2.1 Passion

What are you passionate about and what is important for the public to know about it? Do you want to share the excitement of your own research, or offer your perspective on groundbreaking science research recently reported in the news? Perhaps you believe it is important for students to learn about certain science concepts that are largely ignored in pre-college curricula, for example, the value of plants for human existence (Hershey 2002, 2005; Wandersee and Schussler 2001). Or that concepts and processes inherent in evolution underpin all of human understanding about biological phenomena, and yet some people don't even have an opportunity to hear the word in their high school biology classes. Or that genomics is revolutionizing the way scientists and clinicians think about human disease, and yet most U.S. citizens completed their last science class before this word even entered the scientific lexicon.

Perhaps you think it is important for the public to have a more general understand of what science is, for example, that the goal of science research is to discover new knowledge. The vast majority of the country will never have this experience given that most laboratory learning by students involves demonstrations or the completion of step-by-step procedures to yield predictable outcomes (NRC 2006). Indeed, advocates argue that lab learning is an ideal context for developing an understanding of what science is and how it is done, yet involving students doing experiments doesn't mean they develop this understanding (Bell et al. 2003; Hart et al. 2000; Hofstein and Lunetta 2004; Schwartz et al. 2004).

Maybe you would like students develop specific laboratory skills, or just be able to pick up the newspaper and read about a recent study with a critical and informed eye. Neither students nor the general public has regular opportunities to learn about ongoing

E.L. Dolan *Education Outreach and Public Engagement*,
DOI: 10.1007/978-0-387-77792-4_2 © Springer Science+Business Media, LLC 2008

research (Field and Powell 2001). Even when research findings are reported in the mass media, what is said can differ from the original work (Kua et al. 2004). Even more broadly, you may want the public to consider how basic science leads to the development of new knowledge and understanding or how these findings can be useful for solving problems related to health, agriculture, and the environment. Regardless of your interests, it is essential that they serve as a foundation for your outreach. They will help you set the scope of your work, garner funding, recruit and retain participants, collaborators, and other key stakeholders, evaluate the impacts, and spread the word about what you are doing and why it has meaning.

Perhaps your passion isn't immediately obvious, or your work isn't easily translated or otherwise compelling for a pre-college audience. Then, consider what about your work or your profession would be relevant to their lives or important to remember 20 years from now. The scope of science learning includes subject-specific, interdisciplinary, and multidisciplinary concepts and skills, as well as the process, nature, and history of science. Potential topics include:

- Your own research
- Common misconceptions
- Current events, controversies, and topics of public debate
- Subject-specific, interdisciplinary, and multidisciplinary concepts
- Subject-specific, interdisciplinary, and multidisciplinary technologies and skills
- History, processes, and nature of science

In other words, there is *learning science* (conceptual and theoretical knowledge), *learning about science* (knowledge of the nature and processes of science), and *doing science* (expertise in scientific inquiry and problem solving) (Hodson 1998), also described as conceptual understanding, procedural knowledge, and investigative expertise (Hodson 1996). The field is wide open! For an example of how one scientist translating several aspects of his research to engage high school students, see "Evolving Partnership Between a High School Biology Teacher and an Industrial Researcher" in Sect. 2.1.1.

## 2.1.1  Example of Developing K-12 Outreach and Engagement Activities Based on Current Research

### An Evolving Partnership Between a High School Biology Teacher and an Industrial Researcher

*Contributed by Robert Choy, Exelixis, Inc [rchoy@exelixis.com; www.exelixis.com].*

I first became involved in science education as a graduate student at the University of Washington. I participated in a program organized by the Science Education Partnership at Fred Hutchison Cancer Research Center (FHCRC SEP;

http://www.fhcrc.org/education/sep). This program arranged partnerships between scientists (graduate students, post-doctoral fellows, and professors) and high school biology teachers. During the summer, teachers would visit the scientist's lab to learn about research and design a lesson plan. Later in the school year, the scientist would visit the teacher's classroom and assist with the implementation of the new lesson plan, often loaning equipment, providing supplies, etc. Through this program, I first experienced the joy of teaching and the incomparable feeling of success when a student unambiguously demonstrates that he or she understands something new.

After grad school I moved to the San Francisco to work as a researcher at a biotech company called Exelixis. Rather than cloistering myself in industrial research, even further removed from the Real World than the ivory tower of academia, I sought ways to stay involved with science education. Through the Science and Health Education Partnership at the University of California at San Francisco (the program that served as a model for the FHCRC SEP), I connected with David Lauter, a biology teacher at Washington High School in San Francisco. My partnership with David is entering its 5th year and is constantly being modified and improved. I have been fortunate that my Exelixis supervisors have always supported this endeavor by granting me time off to present two or three lessons in David's classroom each year. Exelixis has also supported my efforts by providing reagents and equipment. My overall goal has been to take something from my own research and present it in a way that is useful and informative to the students. This has always been first and foremost an exercise in communication. I have enjoyed the challenge of translating the complexities of my work in a way that will challenge and excite a young audience. David has played an integral part in this process by making sure that my lessons are presented at a level that is appropriate for his students.

The first set of lessons I co-taught in David's classroom derived from the first project I worked on at Exelixis: using genetic analysis of nematode worms to understand how human pharmaceuticals work at the cellular and molecular level. The class exercises began with the analysis of mutant nematodes with various defects in neurophysiology. We then treated the worms with various drugs in order to examine and measure their responses. This led to a discussion about the similarities and differences between worms and humans and how we could ascertain the cellular mechanism of action of the drugs. After a couple years of running these experiments, the focus of my work at Exelixis changed and I was no longer able to supply David with the worms and reagents. However, he felt the experiments were beneficial to his students and wanted to continue doing them, so I helped connect him with another local worm scientist who volunteered to provide the necessary materials.

I, on the other hand, wanted to bring the focus of my new work, stem cell biology, into the high school classroom. This was a challenge because, while worms can easily be grown in a high school lab setting, culturing stem cells typically requires extensive knowledge, technical expertise, and complex equipment. I enlisted the help of another Exelixis colleague and stem cell expert to help me design hands-on activities using only resources available in a high school lab. In addition to these activities, I also wanted to convey something to the students of the controversy surrounding the ethics of stem cell research. I stumbled upon an entire curriculum from the Genetic Science Learning Center at the University of Utah (http://learn. genetics.utah.edu/) that included an excellent lesson plan for a discussion/debate on the ethical use of stem cells in research. The lesson involves a hypothetical scenario where a couple is considering selling the extra embryos generated during their in vitro fertilization procedure to a biotech company. I have facilitated this lesson numerous times and am always impressed by how vociferously the students argue their positions.

After five years of involvement in this partnership with David, I am embarrassed to admit that we have never conducted any formal assessment. However, David continues to invite me back, demonstrating his belief that my work with his students has value. Informal feedback from the students has also been positive. He also attributes the fact that many of his students go on to major in biology in college at least partially to the influence of my lessons. Most importantly to me, the partnership has grown, evolving with my own changing research interests and with the availability of resources, and yet always producing something engaging and enlightening for the students.

## 2.1.2   Putting Your Passion in Context

Regardless of your passion, it is essential to frame your ideas in the context of the concepts and skills considered important by teachers, schools, districts, and science education policymakers. A good starting point for gaining insight into what is important for students to learn is the local, state, and national science education standards. The national science education standards were published in the mid-1990's with the intent of outlining the essential science concepts and skills students should learn to become scientifically literate citizens (NRC 1996). The standards are organized by discipline and grade level, including what concepts are most appropriately taught together and at what stage of children's development. The American Association for the Advancement of Science (AAAS) spearheaded a complementary effort, titled Project 2061 (http::/www.project2061.org) because of its goal to achieve science literacy nationwide by the year 2061, when Hayley's comet next approaches the earth. Project 2061's *Science for All Americans* (1989) aims to define the knowledge and skills of a scientifically literate individual, and its *Atlas for Science Literacy* (2001) is a large concept map, depicting the connections between concepts, skills, and disciplines and across grade levels.

In early elementary grades, life science standards focus on students' learning about characteristics of organisms, for example, that all organisms have basic needs, including water, air, and an energy source (e.g., food, light, etc.). Students also learn that if these needs aren't met, organisms will not survive. Learning these concepts lays the foundation for understanding the more complex interactions among organisms and their environment that students learn about when they study ecology and evolution. In middle school, these concepts are discussed at the cell, organism, and ecosystem level, while in high school, students consider the molecular underpinnings of a species' success or lack thereof in different habitats.

States, school districts, and even individual schools may have established a more fine-grained set of standards, generally available through state departments of education web site. These standards usually describe what students could and should be learning in different courses at different grades. Additional "scope and sequence" information is often available at the district or state level, describing relationships among scientific ideas as well as what depth of understanding is appropriate at what grade level and what concepts and skills serve as foundations for future learning.

Standards are widely criticized for limiting teachers' instructional choices. Closer examination often reveals that the standards generally make sense and can even be useful for building an argument that particular outreach activities are worthy of class time (see also in Sect. 2.1.3). Standards can be helpful in setting the boundaries of learning appropriate for children of different ages and experience levels, but should not be limiting (Laursen 2006). Rather than dismissing standards outright, use them as guidelines and also consider them seriously in any outreach effort by demonstrating how activities help address standards. Most importantly, consider how to strike a balance between what one is passionate about and what is appropriate for specific classrooms at specific times.

## 2.1.3 Science Education Standards: An Example of the Slippery Slope

What may seem scientifically acceptable in one set of standards can metamorphose in another set, even if the aim was to clarify by using more concrete verbiage. At state, district, and local levels, additional explanations are articulated in the standards themselves and in the complementary guidelines regarding scope and sequence. The text that is carefully crafted and thoroughly vetted by a multitude of science educators, scientists, and science education policymakers, as in the National Science Education Standards (*NRC 1996*), loses its nuances in translation for classroom practice. Concepts are further muddied by a variety of interpretations available in textbooks and the demand for assessments that yield easy-to-quantify data regarding student knowledge.

For example, the national standards include the following description of biological phenomena related to gene regulation:

> Cell functions are regulated. Regulation occurs both through changes in the activity of the functions performed by proteins and through the selective expression of individual genes. This regulation allows cells to respond to their environment and to control and coordinate cell growth and division.... This differentiation is regulated through the expression of different genes (NRC 1996).

The Virginia standards provide numbered lists under headings focused on the molecular, cellular, and organismal level:

> The student will investigate and understand common mechanisms of inheritance and protein synthesis. Key concepts include cell growth and division, gamete formation, cell

specialization, prediction of inheritance of traits…, genetic variation (mutation, recombination, deletions, additions to DNA),… (*VDE 2001*).

The idea of selective expression of genes, including the role of the environment in this process, is present in the national standards but absent at the state level. Perhaps this is because students are able to develop good understandings of enzymes and their functions at a single molecule level, so choices are made to emphasize the process of gene regulation and expression at the level of specific enzymatic processes. Perhaps the complexities and subtleties of how genes are regulated can quickly become overwhelming, considering the full spectrum of processes involved, from the transduction of external signals like light and heat into cells to the signaling within and between cells to detect, transmit, and respond to those signals.

Yet, students hold onto the misconception that genes determine traits, and that the environment, including one's surroundings and behavioral choices, has little effect in determining the expression of genes. Pointing out the absence of any standard that explicitly addresses the interplay between an organism's genotype, behavior, and environment in determining its traits is not to say that this should be addressed explicitly, but rather as an illustration of how the standards are a starting point, not the boundaries for learning. Rather, scientists can use the standards as a tool for considering the appropriateness of any learning activity with regards to its specific topic of study, as well as its depth and breadth.

### 2.1.4 Beyond Your Passion

Consider looking beyond your own discipline. Stepping outside of your specialty can encourage you to look at a phenomenon from a learner's perspective, while still allowing for modeling a scientific approach. As a graduate student, I served as a scientist partner in City Science, an NSF-funded Local Systemic Change Initiative project (http://www.nsf.gov/pubs/1997/nsf97145/projects.htm, accessed 11/19/2007) through the Science & Health Education Partnership (SEP) at the University of California at San Francisco (UCSF; http://biochemistry.ucsf.edu/~sep, accessed 11/19/2007) and the San Francisco Unified School District (http://www.sfusd. edu). Through City Science, experienced teachers and volunteer scientists, usually graduate students or post-doctoral fellows, co-taught week-long professional development sessions for beginning elementary teachers using the district's kit-based curricula (e.g., Insights [http://cse.edc.org/curriculum/insightsElem], Full Option Science System [FOSS; http://www.fossweb.com], etc.).

SFUSD's elementary science curricula crossed the disciplines, including instruction not only in life science, but also in physical, earth, and environmental sciences. Because UCSF is a biomedical institution, many of the scientists were teaching outside of their field of expertise and were often learning from the experienced teachers. Although this scenario was most likely the result of geography rather than program goals, it had the very positive outcomes of dispelling teachers' stereotypes about scientists (e.g., that scientists have all of the answers) and of modeling how a scientist approaches a scientific problem that is new to them. All of the curricula being used in the program were inquiry- or problem-based, meaning that they presented scenarios or issues about which students could make observations, propose testable hypotheses, conduct experiments, and develop new understandings. For example, when a scientist faces a new problem, how does one find relevant information, make observations, critique the information and observations, and use it to develop hypotheses, design and conduct experiments, and develop new explanations?

## 2.1.5   History, Processes, and Nature of Science

Standards not only not only outline content to be learned as it relates to different scientific fields, but also to the history, nature, and processes of science. Although what is meant by the history of science is fairly self-explanatory, discussions about the nature and processes of science are typically the purview of science philosophers and sociologists. The nature of science is defined by what science is and what it is not. For example, science is a human endeavor and scientists are part of a broader scientific community, which has its own values, practices, and mores. The scientific community values "peer review, truthful reporting about the methods and outcomes of investigations, and making public the results of work" (NRC 1996, p. 30). The scientific community is part of a larger society that influences the practices and directions of scientific work. Science is also tentative yet theory-laden. In other words, scientific ideas are based on the development and confirmation of ideas using evidence, and can also be changed or dismissed when new evidence is available. Scientific ideas that have been confirmed time and time again from a number of perspectives and using a variety of tools are less likely to change, and are considered theories.

Learning about the nature of science may seem abstract or so obvious that it seems unnecessary. Yet, proper understanding of science and the scientific enterprise is an essential building block of scientific literacy (Reid and Hodson 1987). Many of the controversies currently under debate across the country are likely because of misunderstandings of what science is and what it is not. It is beyond the scope of this text to describe in any depth the debates regarding embryonic stem cell research or the teaching of evolution alongside creationism or intelligent design in schools. These are just two examples where there appears to be mutually exclusive explanations for the way the world works, one based on faith and the other based on science. Yet, these ways of explaining the world are not mutually exclusive. Rather, faith is based on spirituality and belief and science is based on actual or inferred observations of the natural world. The debate should not be whether faith or science is right, but rather what knowledge is available through faith and what knowledge is available through science.

Similarly, teachers and scientists alike hope that science class is a venue for students to learn about and be engaged in the processes of science. Unfortunately, emphasis on the scientific method often becomes a priority, suggesting that there is one single, universally applicable approach to doing science, and that once one has learned the steps, one will be able to assemble them in order into a coherent whole of doing science. What results are exercises in box checking, where students ask questions, set up experiments that will yield some data, graph those data, and draw some conclusion without careful consideration of the quality of the initial question (e.g., Is there a sound scientific rationale for asking the question? Is the question investigable with the tools available? etc.), whether the investigation is designed to yield data useful for answering the question, whether the data constitute evidence for answering the question, whether graphical or other visual representations of the evidence helps convince the investigators and others of certain conclusions, and what other explanations have already been explored or are hypotheses that serve as the foundation for the design of new investigations.

On the contrary, scientists aim not only to identify and describe causal relationships through their work, but also to identify correlations and solve practical problems (Hodson and Bencze 1998). There is no single algorithm for doing science, and there are few distinct steps (Hodson 1996). Most scientific processes involve developing hypotheses based on existing knowledge and observations, designing investigations to test those hypotheses, collecting and analyzing data, developing explanations based on evidence, developing alternative explanations, and designing investigations or controls to rule out the alternative explanations, but not necessarily in order or in series. Throughout the process, ideas are communicated and developed further, and research questions and methods of data collection and analysis are revised or refined.

In addition, while all science is investigative, not all science is experimental. For example, evolutionary biologists and geologists often don't have the ability to design "experiments" or the luxury to choose controls. Also, some would argue that not all science is hypothesis driven, but instead may be discovery driven or hypothesis generating. For example, there are a number of scientific efforts made possible by the development of high throughput technologies, such as the systematic, community-wide efforts to sequence entire genomes or knock out of all of the genes in an organism's genome. Certainly there is an overarching hypothesis (i.e., if we know the sequence of the genome, then we will gain insights not possible through other means), but most of these efforts are geared to providing data and observations as fodder for generating other hypotheses.

Although scientific content and its history, nature, and processes are discussed separately here, science learning can be enhanced through their integration. For example, by comparing of historical and modern-day scientific ideas, students can learn about how scientific thinking changes with new knowledge and technical capabilities. In addition, learning to make predictions and learning what makes a good prediction is inextricably linked with particular content and theory (Hodson 1996). In other words, it is not possible to make a prediction outside of a particular scientific context, and it is also not possible to evaluate the quality of a prediction without some knowledge about that context. Finally, failure to integrate laboratory learning with the day-to-day concepts and theories students learn about in science class makes it less meaningful and less relevant to learners (NRC 2006).

## 2.2  People

In addition to considering your passion and where it best fits with the goals of K-12 students and teachers, it is important to keep in mind your interpersonal interests and abilities. In particular, with whom do you want to work, what do they need, and what do they have to offer? Do you enjoy working with young children? Are your own children of school age and would you like to offer their classmates a glimpse into the life of a scientist? Do you want to play a role in preparing students for undergraduate coursework? Do you want to develop insights into the skills, abilities, needs, and interests students have when they enroll in the classes you teach? Do you prefer working with adult learners, for example, teachers who can then impact their students' learning year after year? Is your academic year already filled to the brim such that hosting a teacher or a more mature student for a summer lab internship is a better option than school year activities? Answering these questions is the first step in choosing through which of the myriad ways you might collaborate with K-12 teachers and their students.

### 2.2.1  Elementary Students and Teachers

There are strong rationales for working with students of all ages, not only to enhance their learning experiences but also to recruit future scientists. For example, when compared to time spent on science, nearly four times as much is spent on reading and language arts and twice as much on mathematics in the early elementary grades, with only a slight evening of the ratio in late elementary grades (Gruber et al. 2002). This may be the result of an emphasis on standardized testing in reading, writing, and mathematics, or teachers' lack of interest in, enthusiasm about, or preparation to teach science, or uncertainty about their science teaching abilities (Manning et al. 1982; Abell and Roth 1991; Stevens and Wenner 1996). Since students' interest in and enthusiasm about science diminishes during the pre-teen years (Simpson and Oliver 1990; Greenfield 1996; Jovanovic and King 1998), efforts to maximize their interest early on may help students maintain a positive attitude during this tumultuous time.

### 2.2.2  Middle School Students and Teachers

In middle school, equal time is dedicated to each of the subjects, yet teachers are often teaching out of their discipline because of shortages of qualified individuals (Seastrom et al. 2004). Students' interest in science also wanes at this age (Simpson and Oliver 1990; Greenfield 1996; Jovanovic and King 1998). A significant fraction of high school teachers also teach out-of-field, which has a demonstrated impact on student achievement (Darling-Hammond and Hudson 1990; Monk 1994; Goldhaber and Brewer 1997; Ingersoll 1999); only 60% of biology students at the secondary level in 1999–2000 were taught by teachers with a major or minor in biology (Seastrom et al. 2004). Also, eighth graders who expected to have a career in science by age 30 were 1.9 times more likely to complete a baccalaureate degree in life science and 3.4 times more likely to complete a baccalaureate degree in physical science or engineering than those who did not anticipate entering a science career (Tai et al. 2006). This suggests that the middle school years are a prime time for capturing students' interests in a way that still have the potential to affect their pursuit of higher education and, presumably, careers in science.

### 2.2.3 High School Students and Teachers

At the high school level, laboratory learning experiences, which are correlated with positive attitudes toward science and increased science achievement, are not available to all students (Freedman 1997; NRC 2006). When lab learning is available, often it is not integrated into the flow of instruction and does not include time or opportunities for students to reflect on or discuss their work. In addition, the majority of laboratory activities are demonstrations with predictable outcomes. These activities can play a useful role in illustrating concepts or helping students learn techniques. Yet, if demonstration labs are the only laboratory instruction tools in use, students miss an opportunity to experience the excitement of discovery and learn that science is about generating new knowledge. In addition, only a fraction of high school graduates in the U.S. attend college (45% enroll), and an even smaller fraction major in the sciences (17.6% of bachelor's degrees in the U.S. are conferred in the sciences; National Center for Education Statistics 2005). In fact, the last science class most U.S. citizens take is high school biology (Roey et al. 2001).

### 2.2.4 Other Points to Consider

Consider also the logistical constraints of the classroom. Some elementary classes and schools have chosen to adopt a schedule that resembles the regimented subject-by-subject timetable of middle and high schools. Others still have more flexible daily plans, allowing for more time to be spent on a particular classroom activity and integration across topics. Some middle and high schools have extended class time for lab work or hands-on activities, while others are structured as a series of 45-min class periods that meet daily.

At a more personal level, consider whether you prefer to work with younger or older students or adults. Some people find that the enthusiasm of young children is contagious, while others find it exhausting. The developmental maturity of students will also stipulate the depth and breadth of topics that make sense to explore. The best way to get a sense of who is most well-aligned with your interests, expertise, and resources is to explore and listen: talk with teachers, talk with parents, or, if you are a parent, consider your own children and their friends and school mates. Indeed, conversations with teachers are often the most fruitful avenue for determining whether a project, lesson, or activity will work with a group of students. Over the years or even year to year at the secondary level, teachers will work with hundreds of students. Their professional experience as well as their sense of students can be invaluable in guiding the design of K-12 O&E efforts.

## 2.3  Process

Detailed descriptions of how to go about planning and implementing K-12 O&E activities are offered in Chap. 3, while more general factors to consider in advance are offered here. Your interests, time, and resources, as well as those of your K-12 partners, should shape the goals, duration, intensity, and mechanism of your K-12 O&E effort (Fig. 2.1). If you have an interest in working directly with students, you should seek student-focused outreach and engagement opportunities (e.g., mentoring students in science fair projects, leading an after-school science club, etc.; Fig. 2.1). If you are more interested in working with adult learners, seek opportunities to be involved in the professional development of educators. If you have minimal time available during the school day, consider collaborating with teachers to provide curriculum enhancements like experimental materials and kit-based lessons. Finally, if you would like to collaborate with a teacher in working with students, develop an ongoing partnership with a teacher.

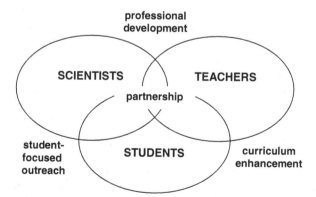

**Fig. 2.1** Mechanisms of K-12 Outreach and Engagement. The involvement of different individuals will guide the type of activity that is possible. For example, kit-based lessons can be co-developed with teachers and either taught by the teacher alone (i.e., curriculum enhancement) or co-taught by the teacher and scientist (i.e., partnership). If the effort only involves scientists and teachers, it is considered professional development

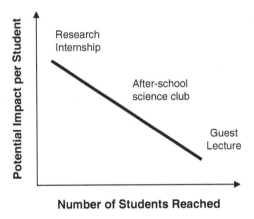

**Fig. 2.2** Hypothetical relationship between the number of students reached and impact per student. Some K-12 O&E activities will reach larger numbers of students, but are less likely to have a long-term effect on students' interest or engagement in science. For example, guest lectures can enhance students' attention during that particular class session but are less likely to influence their choices to pursue a bachelor's degree or career in science. More intensive one-on-one experiences, like research internships, are predicted to be more effective at encouraging students to pursue further education and careers in science

Goal setting is the next step. Although it may seem odd to generate goals after you decide what kind of work is of interest given the time and resources available, keeping these factors in mind from the outset will help ensure that the goals that are established are realistic. Consider and prioritize short, medium, and long-range goals, including changes in (1) knowledge and skills, (2) behavior and attitudes, and (3) status and level of functioning. Changes in knowledge or skills can happen on relatively short timescales, for example, during the course of a lesson or unit. If such activities are repeated several times, many students can be involved. Changes in status, for example, a student's choice to pursue a career in science, will require interactions of longer duration and greater intensity, and thus are less scaleable (Figs. 2.2 and 2.3).

Other factors to consider are the role you would like to play and the scope of your work. For example, do you consider yourself an advocate, a resource, or a partner (Bybee and Morrow 1998)? In other words, do you see yourself speaking out about the value of pre-college science education at meetings of Parent–Teacher Associations or scientific societies? Or developing a longstanding relationship with a local teacher? Do you see yourself working with one or a few students who you hope with pursue careers in science? Or with one or a few teachers who will then reach an exponential number of students? In essence, the choice becomes whether your priority is to interact with many people for a short time or a few people over an extended time, and whether those people are students with whom you will interact directly or teachers who will interact with hundreds or thousands of students over the years to come.

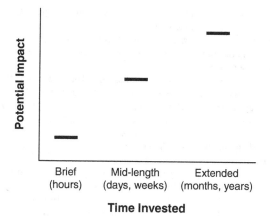

**Fig. 2.3** Hypothetical relationship between time invested and the potential for impact. No causal relationship has been identified between time invested by students, teachers, or scientists in K-12 O&E activities and the impact on those involved, most likely because the logistics of such a study are prohibitive. Such a relationship is hypothesized because there is such a relationship between the duration of teacher professional development and impact on classroom teaching practices (e.g., Garet et al. 2001; Shields et al. 1998; Weiss et al. 1998)

There is no single audience, no single curriculum, no single approach, and no single venue that is "right." Rather, regardless of the people, topics, duration, or intensity of K-12 outreach and engagement activities, the following questions can guide your efforts:

- Who are the key stakeholders and what are ways to ensure they are bought in?
- What do you, your collaborators, and key stakeholders hope to achieve?
- What are ways to ensure that activities, materials, and supplies safe and age-appropriate with respect to students' intellect, interests, and manual dexterity?
- What are ways to ensure the activities are included at a thoughtful place in the curriculum and integrated into the flow of instruction?
- What are ways to ensure that the students see that the activities are worthwhile rather than just busy work, fun stuff to do, or a "day off" of school?
- In what ways will you, your collaborators, and other stakeholders know you have mage progress toward achieving your goals?

# References

Abell SK, Roth M (1991) Coping with constraints of teaching elementary science: A case study of a science enthusiast student teacher. Natl Assoc Res Sci Teach Ann Conf (Lake Geneva, WI, April 7–10, 1991).

American Association for the Advancement of Science (1989) Project 2061: Science for all Americans. Oxford University Press, New York.

American Association for the Advancement of Science (2001) Project 2061: Atlas of science literacy. Oxford University Press, New York.

Bell RL, Blair LM, Crawford BA, Lederman NG (2003) Just do it? Impact of a science apprenticeship program on high school students' understandings of the nature of science and scientific inquiry. J Res Sci Teach 40:487–509.

Bybee R, Morrow C (1998) Improving science education: The role of scientists. Newsl Forum Educ Amer Phys Soc. http://www.spacescience.org/Education/ResourcesForScientists/Workshops/Four-Day/Resources/Articles/1.html. Accessed 3 December 2007.

Darling-Hammond L, Hudson L (1990) Precollege science and mathematics teachers: Supply, demand, and quality. Rev Res Educ 16:223–264.

Field H, Powell P (2001) Public understanding of science versus public understanding of research. Public Understand Sci 10:421–426.

Freedman MP (1997) Relationship among laboratory instruction, attitude toward science, and achievement in science knowledge. J Res Sci Teach 34:343–357.

Garet MS, Porter AC, Desimone L, Birman BF, Yoon KS (2001) What makes professional development effective? Results from a national sample of teachers. Amer Educ Res J 38:915–945.

Goldhaber DD, Brewer DJ (1997) Why don't schools and teachers seem to matter? Assessing the impact of unobservables on education. J Hum Resour 32:505–523.

Greenfield TA (1996) Gender, ethnicity, science achievement, and attitudes. J Res Sci Teach 33:901–933.

Gruber KJ, Wiley SD, Broughman SP, Strizek GA, Burian-Fitzgerald M (2002) Schools and Staffing Survey, 1999–2000: Overview of the Data for Public, Private, Public Charter, and Bureau of Indian Affairs Elementary and Secondary Schools, NCES 2002–313, U.S. Department of Education, National Center for Education Statistics, Washington DC.

Hart C, Mulhall P, Berry A, Loughran J, Gunstone R (2000) What is the purpose of this experiment? Or can students learn something from doing experiments? J Res in Sci Teach 37:655–675.

Hershey DR (2002) Plant blindness: We have met the enemy and he is us. Plant Sci Bull 48:78–84. http://www.botany.org/bsa/psb/2002/psb48-3.html#Plant. Accessed 11 May 2007.

Hershey DR (2005) Plant content in the national science education standards. http://www.action-bioscience.org/education/hershey2.html#Primer. Accessed 23 May 2007.

Hodson D (1996) Laboratory work as scientific method: Three decades of confusion and distortion. J Curric Stud 28:115–135.

Hodson D (1998) Science fiction: The continuing misrepresentation of science in the school curriculum. Curric Stud 6:191–216.

Hodson D, Bencze L (1998) Becoming critical about practical work: changing views and changing practice through action research. Int J Sci Educ 20:683–694.

Hofstein A, Lunetta VN (2004) The laboratory in science education: Foundations for the twenty-first century. Sci Educ 88:28–54.

Ingersoll R (1999) The problem of underqualified teachers in American secondary schools. Educ Res 28:26–37.

Jovanovic J, King SS (1998) Boys and girls in the performance-based science classroom: Who's doing the performing? Amer Educ Res J 35:477–496.

Kua E, Reder M, Grossel M (2004) Science in the news: A study of reporting genomics. Public Understand Sci 13:309–22.

Laursen SL (2006) Getting unstuck: Strategies for escaping the science standards straightjacket. Astron Educ Rev 5:162–177.

Manning P, Esler W, Baird J (1982) What research says: How much elementary science is really being taught? Sci Child 19:40–41.

Monk DH (1994) Subject area preparation of secondary mathematics and science teachers and student achievement. Econ Educ Rev 13:125–145.

National Center for Education Statistics (2005) Digest of education statistics. US Dept Educ, Washington DC. http://nces.ed.gov/programs/digest/d05. Accessed 7 June 2007.

National Research Council (1996) National science education standards. National Academy Press, Washington DC.

National Research Council (2006) America's lab report: Investigations in high school science. In: Committee on High School Science Laboratories: Role and Vision, Singer SR, Hilton ML, Schweingruber HA (Eds.) Board on Science Education, Center for Education. Division of Behavioral and Social Sciences and Education, National Academies Press, Washington DC.

Reid DJ, Hodson D (1987) Science for all: Teaching science in the secondary school. Cassell, London.

Roey S, Caldwell N, Rust K, Blumstein E, Krenzke T, Legum S, Kuhn J, Waksberg M, Haynes J (2001) The 1998 High School Transcript Study Tabulations: Comparative Data on Credits Earned and Demographics for 1998, 1994, 1990, 1987, and 1982 High School Graduates, NCES 2001–498 [Brown, J., Project Officer]. US Dept Educ, Natl Cent Educ Stat, Washington DC.

Seastrom MM, Gruber KJ, Henke R, McGrath DJ, Cohen BA (2004) Qualifications of the public school teacher workforce: Prevalence of out-of-field teaching, 1987–88 to 1999–2000, NCES 2002–603 Revised. US Dept Educ, Natl Cent Educ Stat, Washington DC.

Schwartz RS, Lederman NG, Crawford BA (2004) Developing views of nature of science in an authentic context: An explicit approach to bridging the gap between nature of science and scientific inquiry. Sci Educ 88:610–645.

Shields PM, Marsh JA, Adelman NE (1998) Evaluation of NSF's Statewide Systemic Initiatives (SSI) Program: The SSIs' impacts on classroom practice. SRI, Menlo Park CA.

Simpson RD, Oliver JS (1990) A summary of major influences on attitude toward and achievement in science among adolescent students. Sci Educ 66:1–18.

Stevens C, Wenner G (1996) Elementary preservice teachers' knowledge and beliefs regarding science and mathematics. School Sci Math 96:2–9.

Wandersee JH, Schussler EE (2001) Toward a theory of plant blindness. Plant Sci Bull 47:2–9. http://www.botany.org/plantsciencebulletin/psb-2001-47-1.php#Toward%20a%20Theory%20Oof%20Plant. Accessed 11 May 2007.

Weiss IR, Montgomery DL, Ridgway CJ, Bond SL (1998) Local systemic change through teacher enhancement: Year three cross-site report. Horizon Research, Inc., Chapel Hill, NC.

Virginia Department of Education (2001) Science standards of learning. Commonwealth of Virginia, Richmond VA.

# Chapter 3
# How?

## Developing the Plan and Making It Happen

In general, the process of developing and implementing K-12 O&E activities involves:

- Identifying, as appropriate, the needs of teachers, students, and scientists by listening and observing
- Identifying resources, including materials, technical, and personnel support, that can meet those needs
- Planning and implementing activities that match needs with resources
- Evaluating to determine whether needs are met and to inform future efforts
- Documenting and disseminating the outcomes of the work

This chapter offers advice about initiating, planning, and implementing K-12 O&E activities. Criteria that will help maximize the likelihood of achieving K-12 O&E goals are noted throughout. Also described are a series of concrete examples of projects at different stages of development. Some are just starting while others are well established. Project descriptions are written by the individuals involved, providing glimpses into the different motivations, rationales, and approaches to their work and illustrating the myriad ways they have gone about K-12 O&E.

## 3.1 Identifying Needs

Although the ideas of evaluation and assessment will be more fully developed in Chap. 4, needs assessment is a critical component of K-12 O&E and is most fruitful if conducted prior to program planning or development. Specifically, who are the stakeholders, what do they need, and what resources are available to meet those needs? When initiating a K-12 O&E effort, an informal approach to needs assessment is most appropriate and do-able. Start by listening carefully to teachers, students, and school administrators (e.g., Dolan et al. 2004; Elgin et al. 2005; Lally et al. 2007; Tomanek 2005). What do they want to gain? What are the challenges they face? What types of skills and expertise can they share? What suggestions do they have about how best to meet their needs?

E.L. Dolan *Education Outreach and Public Engagement*,
DOI: 10.1007/978-0-387-77792-4_3 © Springer Science+Business Media, LLC 2008

For example, to inform further development of outreach and partnership programs, we regularly invite high school biology teachers representing a diverse set of schools and students to give us feedback. Several years ago, when the group assembled, they expressed their satisfaction with how our equipment loan and professional development efforts were going, but their students wanted opportunities to collect "real" data. Demonstration labs are essential because they are contexts for students to get hands-on experience with science techniques and they are designed to illustrate specific concepts students are learning through lectures and reading. Yet, much of the fun and excitement in science is experienced during the discovery of new knowledge. Science fairs present opportunities for students to ask and answer their own questions by designing and conducting their own investigations. Yet students only have the opportunity to share their findings with their immediate community of students, schools, and families, rather than the broader scientific community.

We brainstormed about what "real" experiments could be done in a classroom, keeping in mind students' interests, district regulations, and scientist expertise available at the local university. Bacteria and yeast are easy to maintain and manipulate, but safe and sterile preparation, storage, and disposal of microbial growth media can be challenge for many classrooms. Although the group fully supported animal research in science labs, they decided it was inappropriate in classrooms, given the logistics of animal care and district regulations regarding animal use in classrooms. Plants, however, are ideal organisms for students to investigate. They are large enough to be seen and manipulated by young people, inexpensive enough to grow in the scale required in high school classrooms, and hearty enough for student caretakers (Lally et al. 2007).

Our thoughts then turned to identifying the needs of the plant science community. A colleague, Frans Tax at the University of Arizona, noted that the National Science Foundation had established a program, the 2010 Project, the aim of which is to determine the function of all genes in *Arabidopsis* by the year 2010 with the ultimate goal of developing a comprehensive understanding of the biology of flowering plants (http://www.nsf.gov/pubs/2006/nsf06612/nsf06612.htm). *Arabidopsis thaliana*, a member of the mustard family and a relative of the Fast Plants® species (*Brassica rapa*) that is already widely used in K-12 schools, offers two distinct advantages for discovery directed by high school students: it is well-characterized at the molecular level and it is the subject of study by more than 10,000 scientists around the world. Other benefits of using *Arabidopsis* in the classroom include many of the advantages that make it a good model for research, for example: a short life cycle (6–8 weeks), abundant progeny (thousands of seeds per plant), and small size (mature plants are rarely more than 12 in. tall and 8–10 plants can grow in a 3 x 3 in. pot).

Many of the scientists who have received funding from this program have disabled their genes of interest, grown the resulting mutant plants, and looked for any changes in the plant's growth and development, finding no apparent phenotypes (Cutler and McCourt 2005). While it is likely that some genes can serve as "back-ups" for each other, it is also likely that many genes are not used by the plant unless it is responding to environmental stresses like heat, humidity, or pathogens. Plants have had 500 million years of evolution to adapt to every biome on Earth. The stationary nature of their existence would suggest an arsenal of genes that allow the plant to accommodate changes in their environment. Frans proposed that growing mutant plants under stress conditions would allow a more comprehensive analysis of gene function. This is the basis of the Partnership for Research and Education in Plants (PREP; http://www. prep.biotech.vt.edu/; NSF DBI-9975808; NSF MCB-0418946; grant number R25 RR08529 from the National Institutes of Health National Center for Research Resources Science Education Partnership Award program). PREP provides genuine research experiences to high school students and teachers while helping scientists to discover the function of previously uncharacterized plant genes. High school students conduct experiments on mutant lines of *Arabidopsis* under a variety of stress conditions and then analyze their phenotypes, reporting their findings back to their partner scientists (Lally et al. 2007).

The example of how PREP got started illustrates how the needs of all stakeholders were identified and then served as a foundation for program development. In this situation, stakeholders included high school biology students, their teachers, and plant scientists:

The students and teachers:

- Wanted to do "real" experiments that are of interest to scientists in the context of biology class
- Had space, hands, minds, and classroom know-how

The scientists:

- Wanted to characterize thousands of genes in the plant *Arabidopsis thaliana*, and broaden the impact of their research
- Had seeds and knowledge about experimental design, genetics, genomics, and biotechnology

The articulation and documentation of these needs, and design of PREP to respond to them, served as the basis for garnering extramural funds from NSF and the Science Education Partnership Award program of the National Institutes of Health National Center for Research Resources. For other examples and advice regarding assessing needs, see the W. K. Kellogg Foundation's Evaluation Handbook and the descriptions of specific K-12 O&E efforts featured in this text, including the description of *Biotech-in-a-Box* (below).

### 3.1.1  Example of Identifying and Responding to Needs

#### Biotech-in-a-Box

Contributed by Kristi DeCourcy, Fralin Center, Virginia Tech [decourcy@vt.edu; additional information available at: http://www.biotech.vt.edu/outreach/biotech_box.html].

The Fralin Center at Virginia Tech began its outreach effort in 1992 with biotechnology-related workshops for Virginia teachers. When participating teachers were asked what other things the Center could do to help them bring biotechnology to their classrooms, they replied that they could not afford the equipment and materials to allow their students to do biotech experiments. Biotech-in-a-Box began in 1994 in response to those teachers' requests. The goal of the program is to enable students themselves to do the experiments hands-on, instead of just reading about or seeing the procedures demonstrated.

In 1994, one DNA Biotechnology Kit circulated to seven schools, serving 300 students. From that start, the program has grown to over 30 kits serving over ten thousand students annually. Cases of equipment and materials are shipped to the school for two-week loans. The Center covers all costs, including round-trip shipping. Due to finite resources, the program cannot serve all Virginia schools at all grade levels, so Biotech-in-a-Box targets high school and community college students. By this time, students have the background to understand the science and are likely to derive the most benefit from hands-on experiments. Since many students (including me!) first get excited about science in middle school, we decided to make some materials available at lower grade levels, but the middle school effort is a small part of the program. The program serves both public and private schools, as we decided early not to discriminate between schools no matter what their funding or philosophical leanings- we are targeting all Virginians. In addition, equipment and materials are made available for summer programs and for teacher in-service training, both of which provide benefits to the program. Fralin personnel reach limited numbers of teachers for training during the year, but when teachers use the equipment to train other teachers, our impact is amplified. Using the equipment for summer programs maximizes equipment use, as there are few classroom loans during the summer.

All kits include equipment and all materials except for a few grocery store items, thus reducing the time and effort that teachers have to expend to prep for the experiments. Kit manuals include background information, detailed information on kit use, student materials, and scenarios for classroom use. The manuals are sent to the instructor at the beginning of the semester in which the kit will be used so the teacher can best integrate labs into what they are teaching. The kits currently offered are:

- DNA biotechnology-students conduct electrophoresis of DNA in the context of a crime scene or ethics scenario and/or to learn lab skills (e.g., when students are enrolled in technician preparation programs).
- Protein biotechnology-students conduct electrophoresis of proteins in the context of experiments exploring gene expression, evolution, etc.
- Column chromatography-students separate molecules in a mixture by size, charge, and polarity, learning about molecular characteristics.
- Introduction to immunology-students spread a "disease" within the class, then use an immunoassay to determine who is infected and identify who was "patient zero."
- Thermal cyclers may be borrowed by instructors wishing to do polymerase chain reaction experiments in the classroom using materials they purchase separately.

The Fralin Center has an endowment that covers all of the costs of the program, though some grant funding has provided additional equipment. Costs include round-trip shipping, expendable materials, equipment replacement, and salaries of the work-study students who provide most of the labor. Initial equipment set-up costs for a single kit range from less than $200 (Immunology Kit) to around $4,000 (DNA Biotech Kit, the most "equipment-heavy" kit). Replacement equipment costs average 10% annually. The largest annual costs are shipping and expendable materials. Since Biotech-in-a-Box serves all of Virginia, almost all of the kits must be shipped (some local teachers pick up the materials at Fralin). Round-trip shipping by 2-day freight service costs between $15 (Chromatography Kit) and $100 (DNA kit), depending on distance shipped and weight of the kit. Annual expenses for expendable materials vary as well. For an individual kit, average annual expenses are: DNA-$1,000; chromatography-$200; protein electrophoresis-$900; and immunology- $200.

As administrator of the program, I do the scheduling of the kits, provide support to teachers, prepare kit checklists from which the work-study students pack the kits, and prepare most of the lab-made expendables, such as digested DNA and buffers. I work on the program about 20 h at the beginning of each semester, then 1–2 h a week during the academic year. The two student workers unpack returning kits, check all equipment, and repack the kits, working a combined total of 15–20 h/week.

Initially, information about Biotech-in-a-Box was disseminated at workshops and through a newsletter sent to Virginia educators. Both still bring new kit borrowers, but more new borrowers now learn of the program through word-of-mouth. Each year, a few teachers drop from the program because they retire, change schools, or otherwise move on, but new teachers sign on each year as well. One indication that the program is successful in assisting teachers in their classrooms is that many teachers borrow equipment year after year. Five of the original seven borrowers from 1994 are *still* borrowing the kits every year. Feedback from the borrowers is universally positive, and many state that they could not bring these experiences to their students without Biotech-in-a-Box. One of the original 1994 borrowers, who teaches in a small town in far southwest Virginia, writes a thank-you note after every loan, thanking us "for helping us show our students what else is out there in the world!"

Although I strive to make Biotech-in-a-Box as user-friendly as possible, I have established some restrictions that make the program easier to run. I plan the loan calendars, which specify the loan periods for each kit, around the Virginia Tech academic calendar and student breaks, so that work-study students are available for most of the kit packing. Kit materials are standardized as much as possible to keep packing simple. For example, pipettes are packaged in bags of ten, so any kit that needs pipettes gets one bag per class, even though that may be more than they need for the experiment.

One facet of the program seems to be key to its success and to ever-increasing demand by Virginia educators: *flexibility*. Kit borrowers themselves are free to decide if, when, and how the kits fit into their curriculum. For example, a teacher may borrow the DNA Biotech Kit for one class of 15 marine biology students to teach how scientists discriminate between different marine organisms, whereas a second teacher may borrow the same kit to use with 150 Biology I students to simulate a crime for their students to solve using DNA fingerprinting. The "IF" is key. I have found repeatedly that, when administrators like principals or science supervisors mandate when their teachers use the kits (i.e., not "if"), the borrowing teacher will use the kit for a single year and never again.

## 3.2   Matching Needs with Resources

Developing a logic model is an ideal way to guide the planning of K-12 O&E activities. Logic models help the individuals involved align project activities with goals and ensure that the identified needs and available resources fit what the project is designed to achieve. Specifically, logic models are visual organizers for aligning the goals, objectives, activities, and impacts of a program. Usually organized as flow diagrams, logic models include inputs (what goes into the program, e.g., types of participants, available resources, even the program's social, cultural, and political environment), activities or strategies (what happens during the program and how it happens), outputs (what comes out of the program, e.g., program events, educational materials developed, numbers of participants, etc.), and outcomes (what happens as a result of the outputs, from short term effects like learning of specific concepts or skills to long term effects like changes in behavior or in social or economic conditions). Long-term outcomes often are related to overall goals and are usually the underlying motivation for the activities.

There are a number of freely available resources that provide guidance on developing logic models (e.g., W. K. Kellogg Foundation web site under Knowledgebase, Toolkits, Evaluation; Program Development and Evaluation through the University of Wisconsin Cooperative Extension, http://www.uwex.edu/ces/pdande; Frechtling 2002). As an example, the personnel associated with Biotech-in-a-Box (described in Sect. 3.1.1.1) developed a detailed logic model to identify more formal ways to evaluate the impact of the program (Table 3.1). In the process of generating the logic model, goals and outcomes of the program were carefully defined and prioritized. In addition, the logic model was useful in determining appropriate evaluation strategies given available time and resources, and in identifying data that were already collected that could be useful in unanticipated ways (e.g., determining WHEN teachers borrowed the kits could provide hints about their learning objectives, which could then be examined more directly and systematically).

When developing a logic model for a new effort, a good place to start is at the end: what goals do you want to achieve? These goals are typically represented in the longer-term outcomes, and serve as a starting point for planning the intermediate objectives that lead to those goals. Similarly, short-term objectives can be identified that result in the intermediate objectives. Short-, medium-, and long-term activities can then be developed with the aim of addressing goals and objectives over time. Of course, as activities are developed, piloted, and revised and goals and objectives change, the logic model can be refined or reformulated.

**Table 3.1** Biotech-in-a-Box logic model

| Inputs | Processes | Outputs | Outcomes |
|---|---|---|---|
| Five different Biotech in a Box kits (30 total), which include:<br>• Teacher manuals<br>• Template student handouts<br>• Equipment<br>• Consumable materials<br>• Scenarios<br>• Lessons generated by teachers<br><br>Program director<br>Support from Center Director, including endowment funds<br>Outreach director<br>Two undergraduate students who pack and unpack the kits<br>Fedex for kit shipping<br>Secondary and community college teachers who borrow kits<br>Secondary and community college students who use kits | The Biotech in a Box program is the longest standing public education effort of the Fralin Biotechnology Center (initiated in 1994).<br><br>Through this program, high school and community college teachers receive equipment and materials to guide their students in conducting hands-on, activity-based lessons, problems, and/or units in biotechnology.<br><br>Biotechnology is defined as the products of life science, including DNA fingerprinting, genetically-modified foods, and disease testing materials, as well as methods for examining the characteristics of molecules (e.g., size, charge, etc.). | Number of:<br>• Teachers borrowing<br>• Students participating<br>• Schools participating<br>• Kits loaned<br>• Lessons generated by teachers<br><br>Types of classes where kits are used (e.g., biology, chemistry, biotechnology)<br><br>Level of classes where kits are used (e.g., high school biology, AP biology, introductory college biology, etc.)<br><br>Media contacts<br>• Number of press releases<br>• Number of press clippings<br>• Number of phone calls | Students develop an understanding of principles in: genetics, cell biology, chemistry, epidemiology, immunology, evolution, and forensics<br><br>Students develop understanding of practical applications of biology and chemistry<br><br>Student awareness of biotechnology as a career option<br><br>Provide useful and necessary resources for high school and community college biology education<br><br>Teachers integrate hands-on and lab-based lessons into their curricula<br><br>Positive and more informed public perception of biotechnology<br><br>Students think science is interesting and fun<br><br>Positive public perception of Virginia Tech<br><br>More informed decision-making regarding biotechnology by the public<br><br>Student awareness of Virginia Tech as a college option |

## 3.3  Choosing the Approach

Once you have initiated a conversation with a teacher or school administrator and discussed how best to match your needs and resources to a mutually beneficial end, what do you actually do?

### 3.3.1  Join Existing Programs

First, consider joining an existing project with an established network of participating schools so that you can spend your time working with students and teachers rather than designing and administering a program (Lally et al. 2007). For example, the Botanical Society of America has established an online plant science mentorship program, PlantingScience (http://www.plantingscience.org). In this program, students can design and conduct hands-on investigations with in-class guidance from their teachers and web-mediated mentorship from plant scientists. Schools themselves may have existing mechanisms for scientists to get involved, for example, through science fairs or volunteer programs. An example of a program seeking interested scientists, see "Central Virginia Governor's School Electron Microscopy Center: Wow Moments for K-12" in Sect. 3.3.2.

A number of organizations, especially land grant universities with well-developed outreach and extension programming, have either physical or virtual clearinghouses of projects and resources for working with K-12 schools, for example, University of Arizona's Science and Mathematics Education Center and Virginia Tech's K-12 Science, Technology, Engineering, and Mathematics Education Outreach Initiative. These sites serve as portals to resources, projects, and personnel with special interest and expertise in working with K-12 audiences. Professional societies (e.g., American Society for Microbiology, etc.) and grants-making organizations (e.g., National Science Foundation, National Institutes of Health, Howard Hughes Medical Institute, etc.) often make their K-12 education resources available through their own web sites or the sites of grantees. For example, NSF's Plant Genome Research Program has supported the development of K-12 plant science education resources and the Plant Genome Research Outreach Portal that makes these resources accessible (http://www.plantgdb.org/PGROP/pgrop.php). Examples of funding agencies and professional organizations to explore are outlined in Table 3.2.

Even if you eventually decide to go your own direction, participating in an existing effort can save you time, effort, and energy as you "learn the ropes" of K-12 O&E. For first-hand accounts from individuals who started their own collaborations after making connections through existing programs, see the examples "Margaret Beeks Elementary Outreach" in Sect. 3.3.3 and "Grassroots Research Collaborations: Blurring the boundaries between classrooms and laboratories" in Sect. 3.3.6.

**Table 3.2** Other venues for identifying existing K-12 education and outreach resources

**Grant-making organizations**

| | | |
|---|---|---|
| Howard Hughes Medical Institute | | http://www.hhmi.org/research/professors; http://www.hhmi.org/resources |
| National Institutes of Health (R25) | National Center for Research Resources | http://grants.nih.gov/grants/guide/pa-files/PAR-06-549.html; http://www.ncrrsepa.org |
| | National Institute on Alcohol Abuse and Alcoholism | http://grants2.nih.gov/grants/guide/pa-files/PAR-07-001.html |
| Science Education Partnership Award | National Institute of Allergy and Infectious Diseases | http://grants.nih.gov/grants/guide/pa-files/PAR-08-003.html |
| | National Institute on Drug Abuse | http://grants2.nih.gov/grants/guide/pa-files/PAR-06-518.html |
| | National Heart, Lung, and Blood Institute | http://grants.nih.gov/grants/guide/rfa-files/RFA-HL-07-013.html |
| National Science Foundation | Elementary, Secondary and Informal Education | http://www.nsf.gov/od/lpa/news/publicat/nsf04009/ehr/esie.htm |
| | Research, Evaluation and Communication | http://www.nsf.gov/od/lpa/news/publicat/nsf04009/ehr/rec.htm |
| | Graduate Teaching Fellows in K-12 Education | http://www.nsf.gov/funding/pgm_summ.jsp?pims_id=5472 |
| | Math and Science Partnership | http://www.nsf.gov/funding/pgm_summ.jsp?pims_id=5756 |
| The Wellcome Trust | | http://www.wellcome.ac.uk/funding/publicengagement |

**Other professional organizations**

| | |
|---|---|
| American Society for Biochemistry and Molecular Biology | http://www.asbmb.org/ |
| American Society for Cell Biology | http://www.ascb.org/index.cfm?navid=6; http://www.lifescied.org |
| American Society of Human Genetics | http://genetics.faseb.org/genetics/ashg/ashgmenu.htm; http://www.genednet.org/ |
| American Society of Plant Biologists | http://www.aspb.org/education |
| Botanical Society of America | http://www.botany.org/outreach |
| Ecological Society of America | http://www.esa.org/teaching_learning/educatorResources.php |
| National Association of Biology Teachers | http://www.nabt.org |
| National Science Teachers Association | http://www.nsta.org |
| Society for Developmental Biology | http://www.sdbonline.org/ |
| Society for Neuroscience | http://www.sfn.org/ |

### 3.3.2  Example of an Existing Program in Search of Interested Scientists

**Central Virginia Governor's School Electron Microscopy Center: Wow Moments for K-12**

Contributed by Cheryl A. Lindeman, Partnership Coordinator and Microscopy Center Laboratory Manager [clindema@cvgs.k12.va.us; additional information available at: http://www.cvgs.k12.va.us/CURRIC/SRSEM/emlab/index.htm].

Central Virginia Governor's School in Lynchburg, VA was established in 1988 to create unique learning experiences for gifted students in science and mathematics. The Electron Microscopy Laboratory started in 1989 with a donation of a transmission electron microscope from the University of Dayton. Because of my experience with microscopy in undergraduate and graduate school, I became the microscope point-of-contact. I decided to develop a microscopy-based lab curriculum, which was integrated into our senior technology seminar course, so that our students could take full advantage of this rare (at least, at the high school level!) piece of equipment. The school was also able to acquire a scanning electron microscope (SEM), but both scopes were decommissioned in 2006. Spurred by demand from students, their families, and the faculty, the school along with the CVGS Foundation, parents, alums and local school divisions and businesses joined forces and resources to acquire a brand new Hitachi N3400 Type II Variable Pressure SEM.

The goal of Electron Microscopy Center is to provide unique learning opportunities for CVGS students using SEM technology by:

- Presenting better-than-the-naked-eye images that can be used for classroom instruction across grade levels of interesting living and nonliving materials so that regional teachers can access them via the SEM's unique web interface technology.
- Developing the Center into a regional or national electron microscopy teaching center for pre-college educators.

The operating budget of the laboratory is supported by the CVGS Foundation, teacher-authored grants, and school funds. Some of these funds enabled the purchase of a Taipan Server to allow students in our area to view specimens under the scope while in their classrooms using simple and widely available web technology. The server was installed in fall 2007 and area teachers are in the process of getting the connectivity needed to access and download images. CVGS students are selecting a wide variety of specimens, learning how to prepare and analyze the images and developing appropriate "WOW" lessons that can be used to teach the Virginia Standards of Learning. Also, minigrants from the U.S. Department of Energy through the Pennsylvania State University (Subaward No. 2579-CVGS-DOE-4423 and Subaward No. 3492-CVGSST-DOE-4423) provides funds to explore the gamma irradiation effects on living and material science specimens. By Summer 2008, we expect to have other partnerships established between CVGS students and faculty and researchers across Virginia who are interested in using SEM to enhance their understanding of model organisms or taxonomical projects. Thus, students are asking and answering their own research questions, developing instructional resources for other students, and even preparing to collaborate with practicing scientists in their research. Wow!

### 3.3.3 Example of Initiating an Independent Project after Involvement in an Established Effort

*Margaret Beeks Elementary Outreach*

*Contributed by Glenda Gillapsy, Biochemistry, Virginia Tech [gillaspy@vt.edu; additional information available at: http://www.biochem.vt.edu/gillaspy/mbeeksfast-plants.htm].*

My collaboration with elementary science classes was stimulated by my professional activities as a scientist partner in the Partnership for Research and Education in Plants (PREP). When I started working with PREP, I was looking for ways to increase students' critical thinking in my own large enrollment, undergraduate biochemistry courses, and was impressed by the ability of high school teachers to bring "real" scientific experimentation to the classroom. As a parent of elementary school children, I was interested in volunteering my time at my children's school, and helping provide science resources for teachers. I was also beginning to hear this phrase at home "Mama, can we do a science experiment?" I had been in the school providing hands-on demonstrations using plants and microscopes, but guiding students in conducting experiments seemed like a much more relevant and authentic science experience for them. Thus, the goal of my collaboration with elementary school teachers is give their students a chance to ask and answer their own questions about biology by providing the biological materials and a general framework. Since one major limitation for teachers is the time and expense required to utilize living organisms, I bring plants and growth materials to the classroom. I also provide knowledge on plant stress physiology to help guide the experimentation. I also give graduate students and post-doctoral fellows from my lab the option to help out with a class and provide additional expertise.

In general, I approach teachers to see if they are interested in collaborating, but some teachers hear about past activities and contact me to express their interest. I discuss the elements of the project with the teacher, who then tailors the experiments to their specific learning goals. In Virginia, these learning goals are defined by the state, however, I have found that our set-up can be adapted to address many different learning objectives. For example, the project meets the needs of Virginia third graders who are expected to conduct investigations, observe living organisms, and understand how environments support plants. It can also be easily adapted for fourth graders who need to investigate and understand basic plant anatomy and life processes. Both fourth and fifth grade standards focus on growth in understanding the nature of science and data analysis, so the project can be modified to include more complex data analyses.

I usually visit the classroom three or four times for about an hour each visit to introduce and start the project, help collect data, and help interpret data and draw conclusions. During the first visit, groups of students plant their seeds and we discuss plant growth and stress. I have used *Arabidopsis thaliana* seeds from my own research program as well as Fast Plants (*Brassica rapa*; http://www.fastplants.org/) and other seed obtained from Lehle Seed. Students then brainstorm to choose a specific stress and are given help in designing an appropriate experiment in which to test the effects of this particular stress. In one third grade class, a group of students wanted to examine the effect of wind stress, so they decided to use a box fan to simulate wind. The second classroom visit usually involves the application of the chosen stresses, during which we discuss the concept of experimental controls. Since many stresses have short-term effects, I usually visit the class the day following stress application, where we discuss their observations. After this point, students often make observations and measurements on their own, recording their data. The final visit involves drawing conclusions from the data. This is an ideal time to integrate mathematical concepts like estimates and averages, as well as provide a forum for group presentations of the results. If photographs are taken, a quick web site can be put together to allow students to show their results to their families via the internet.

A strength of this project is that it can be adapted to any organism that can be cultivated in the classroom. For example, along with other scientists who are parents, I am exploring zebrafish as an alternative organism for inquiry-based elementary science experiments. Since the cost of both the plant and zebrafish systems is minimal, no external funding has been pursued for this project. Although the impact of this project is difficult to determine at this point, new teachers have expressed interest in participating and other parents have expressed interest in volunteering their time in the project. Perhaps the best evaluation is from the students, who are enthusiastic and engaged and always ask to do more experiments!

### 3.3.4  Curriculum Enhancement

As in the example from Margaret Beeks Elementary, there is significant appeal to enhancing the existing curriculum in schools, perhaps by adding hands-on activities or engaging students in designing and conducting experiments. Again, take advantage of existing lessons, units, or materials to save time, energy, and effort, and to ensure that they are classroom-tested and of good quality.

> MYTH: The problem with science education is a lack of good curriculum and therefore we must develop it (James M. Bower, University of Texas Health Sciences Center at San Antonio).

The development of high quality curriculum is time-consuming and expensive, requiring significant breadth and depth of expertise in teaching, learning, educational

**Table 3.3** Example resources for K-12 science education

| | Curricular or project resources | Topic | Web site |
|---|---|---|---|
| Elementary | Full Option Science System (FOSS) | Multiple | http://www.fossweb.com |
| | Great Explorations in Math and Science (GEMS) | Multiple | http://www.lawrencehallofscience.org/gems/ |
| | Insights | Multiple | http://cse.edc.org/curriculum/insightsElem |
| Secondary | Dolan DNA Learning Center; DNA Interactive | Genetics, biotechnology | http://www.dnalc.org; http://www.dnai.org/ |
| | Genetic Science Learning Center | Genetics | http://learn.genetics.utah.edu |
| | Howard Hughes Medical Institute Biointeractive | Life science | http://www.hhmi.org/biointeractive/vlabs/ |
| | National Institutes of Health Curriculum Supplements | Biomedical science | http://science.education.nih.gov/supplements |
| | Virtual Plant Biotechnology and Genomics Laboratory© | Biotechnology | http://ppge.ucdavis.edu/Software/software.cfm |
| Cross Grade Level | BioEdOnline | Life science | http://www.bioedonline.org/ |
| | Biological Sciences Curriculum Study (BSCS) | Life science | http://www.bscs.org/ |
| | Fast Plants® | Life science | http://www.fastplants.org |
| | GLOBE Program | Ecology/conservation | http://www.globe.gov |
| | Monarch Watch | Ecology/conservation | http://www.monarchwatch.org/ |
| | Project WILD | Wildlife conservation | http://www.projectwild.org/ |
| | Stream Team | Ecology | http://www.stream.fs.fed.us/ |
| Digital Libraries and Portals | American Physiological Society (APS) Archive | Physiology | http://www.apsarchive.org |
| | BioSciEdNet BEN Portal | Life Science | http://www.biosciednet.org/portal/ |
| | Microbe Library | Microbiology | http://www.microbelibrary.org/ |
| | Plant Genome Research Outreach Portal | Plants, biotechnology | http://www.plantgdb.org/PGROP/pgrop.php |
| | Science Education Partnership Awards | Multiple | http://www.ncrrsepa.org |

research, and science knowledge. A wealth of "tried and true" science curricula have been developed that span the K-12 curriculum (see Table 3.3 for examples). Some can be used to illustrate science concepts, while others involve students in the process of scientific inquiry. Although these materials may not directly fit the learning objectives you have in mind, you and your school colleagues can alter them to fit your needs.

For the most part, teachers are interested in identifying anything that helps their students learn. Teachers regularly adapt learning materials to address a gap in their existing curriculum or improve on the curriculum already in use, especially if the materials are low cost, scaleable (in some states, high school teachers have 175 students or more), and flexible (easily integrated into existing curriculum, e.g., Elgin et al. 2005; Evans et al. 2001). Lessons, activities, and units are more likely to be used by teachers repeatedly if they:

- Fit into the regular time allotted for science learning (e.g., 45–50 min at the secondary level, more or less at the elementary level),
- Include suggested stopping points (i.e., if the activity takes longer than the allotted time, either by design or because of an unexpected fire drill, where is a good point to pause until the next day?),
- Are aligned with the standards (i.e., include a chart outlining what standards, skills, or knowledge students are likely to learn at which points in the lesson), and
- Are flexible and responsive (i.e., Is there a way for teachers to give feedback? See how other teachers use the materials? Use the materials in a number of ways to teach different concepts or skills depending on their students or the particular class?).

As in several examples described here, scientists can be involved directly in teaching the lessons or leading the activities, with the help and guidance of a collaborating teacher. If a teacher thinks that a particular curriculum enhancement is worthwhile, he or she will make special efforts to integrate it into their curriculum year after year, even when a scientist is not available as in Sect. 2.1.1.1 or 3.1.1.1.

## *3.3.5   Student–Teacher–Scientist Partnerships*

Although any classroom-based activities could enhance school curricula, partnerships are distinguished here because they are designed from the outset to involve mutual contribution and benefit (Fig. 3.1). Research collaborations, through which students and teachers contribute to the ongoing research of their partner scientists, are one strategy for partnership. In these particular relationships, the benefits realized by the scientist go beyond the acquisition of teaching and communication skills and the eventual improvement of public science literacy to include the direct involvement of students and teachers in the scientists' research. Such partnerships can generate new directions for research as students' and teachers' creative and unpredictable approach to scientific problems offers a fresh

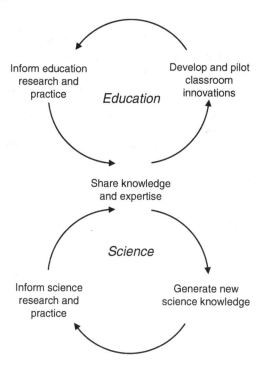

**Fig. 3.1** Complementarity of science and education. Educators and scientists participate in overlapping cycles of scientific and educational expertise by generating new educational and scientific knowledge, respectively. Through partnerships, this new knowledge has the potential to enhance the pedagogical and scientific expertise of each partner

perspective. Students may ask questions that lead scientific investigations in new directions, resulting in unexpected scientific findings. For example, high school students working in the University of Minnesota's monarch butterfly research lab spend several weeks working on research projects directed by undergraduate and graduate students and faculty members in the laboratory. After this introduction, the students are encouraged to develop and test their own hypotheses, often culminating in peer-reviewed publications. In one case, the work spearheaded a completely new line of inquiry for the laboratory: the effects of temperature on monarch larval coloration (Solensky and Larkin 2003). Another high school student carried out a detailed study of the sub-lethal impacts on monarch larvae of an insecticide commonly used for mosquito control; this part of the study would not have been conducted without the student's motivation and effort (Oberhauser et al. 2006).

Such partnerships are more likely to succeed if everyone is mindful that:

- Materials and supplies must be appropriate for pre-college classrooms (e.g., safe, inexpensive, scaleable).
- Publishable data are rarely collected the first time anyone conducts an experiment.

- Students have little experience with the tedium of collecting data carefully and repeating experiments, or with the disappointment of negative results.
- Students (and even teachers!) may have little experience with the "messiness" of science, especially the ambiguity that is typical of results from a first round of experimentation and the unpredictability that is inherent in working with live samples.

Research collaborations are likely to benefit from more day-to-day involvement by scientists who can model how they go about making observations, considering what data constitute evidence, and making inferences (Grady and Dolan submitted). Alternatively, teachers can develop a more in-depth understanding of the research through summer research internships. For examples of student–teacher–partnerships, see "Grassroots Research Collaborations: Blurring the Boundaries Between Classrooms and Laboratories" in Sect. 3.3.6 and "Living among the Extremophiles: Collaborative Investigations of Cyanobacteria as a Context for Learning Chemistry and Biology" in Sect. 3.3.7.

Research collaborations are only one form of partnership. Other formats include after-school science clubs for students who are co-mentored by teachers and scientists (e.g., the Triad Alliance for Gender Equitable Teaching at the Science and Health Education Partnership at University of California at San Francisco http://biochemistry.ucsf.edu/~sep/triad.html). Other program structures engage additional partners, like families, including parents, grandparents, and siblings. For example, University of Maryland's Be Active Kids! Program (http://www.sph.umd.edu/BeActiveKids/) hosts regular Family Science Nights, during which children lead their families through activities and experiments regarding exercise, cardiovascular endurance, caloric balance, and muscle strength. Other partnership models are listed or described in depth elsewhere (e.g., Bybee and Morrow 1998).

## 3.3.6 Example of Initiating an Independent Partnership after Involvement in an Established One

### Grassroots Research Collaborations: Blurring the Boundaries Between Classrooms and Laboratories

*Contributed by John Kowalski, Roanoke Valley Governor's School, Dorothea Tholl, Biological Sciences, Virginia Tech, and Boris Vinatzer, Plant Pathology, Physiology and Weed Science, Virginia Tech.*

The Roanoke Valley Governor's School for Science and Technology is a half-day program for exceptional students from six school districts in the Roanoke area. Students attend the Governor's School for a half-day each day, taking different mathematics, science, and elective courses each year. Student participation in research is an important component of the Governor's School experience. There is an emphasis on hands-on lab experiences in both the mathematics and science courses. All students also participate in an elective course that serves as a vehicle for individual and group research projects. Aspects of experimental design, data collection, and data analysis and interpretation are discussed in both the regular and elective courses.

Students in the biology courses have participated for several years in the Partnership for Research and Education in Plants (PREP). Participation in PREP has provided students with a variety of experiences that fit into the school's research emphasis:

- Students participate in experimental work that does not have a predicted outcome. This approach complements the typical high school labs that demonstrate concepts and build skills but have known outcomes.
- Students come to understand that they can participate in "real-world" research and make contributions to our understanding of current research questions.
- Students interact with research scientists while learning about current research and the people who carry out that research. Students leave the experience with a greater understanding of modern research and, for many of them, a new found interest in becoming scientists.

John Kowalski, a biology teacher at the school, is currently a PREP Fellow. PREP Fellowships are awarded to a select group of experienced teachers across the country who can use the fellowship to achieve their own goals and inform further development of the program. Kowalski's goals for his fellowship are to:

- Work in local labs to gain experience with current research questions and techniques that can be applied to the model plant *Arabidopsis thaliana*,
- Develop and pilot protocols that can be used in high school settings, and
- Develop relationships with local scientists that lead to their active participation in collaborations with high school students.

This year, Kowalski is working with two Virginia Tech scientists, Dorothea Tholl and Boris Vinatzer, to bring their research into the high school classroom. These experiences provide students with the opportunity to interact with research scientists and learn about current areas of investigation that may lead to the control of important agricultural pests and pathogens. These projects are particularly appropriate for pre-college students since they illustrate how research can have direct practical applications. Students will interact with the researchers using a variety of means including face-to-face meetings, video clips prepared by the researchers, and video conferences. The student groups will prepare summaries of their work using PowerPoint for presentation to the researchers. In addition, the protocols and lessons developed during these collaborations will be classroom-tested, setting the stage for involvement of students and teachers at other schools.

*Collaboration with Boris Vinatzer.* One of Kowalski's biology classes is participating in a research project with Boris Vinatzer, whose research includes the identification and characterization of bacterial plant pathogens with the aim of developing approaches for their bio-control using bacteria. Strains of the bacterium *Pseudomonas syringae* are important pathogens of crop plants such as tomatoes and tobacco. Non-pathogenic strains of *P. syringae*, however, might serve as bio-control agents by competing with pathogenic bacteria, thus protecting the plant.

While working with Vinatzer, students collected wild-growing plant samples from the area around the school and cultured them to obtain any *P. syringae* bacteria growing on them. Future groups of students will infect tobacco plants with their isolated bacteria to determine if they are pathogenic or non-pathogenic. Finally, students will use polymerase chain reaction (PCR) technology to identify pathogenic and non-pathogenic strains of bacteria based on known sections of their DNA that are associated with pathogenicity. Students will have the opportunity to learn a variety of concepts and skills including skills including sterile technique, culturing and identifying bacteria, PCR, evolution, and bioinformatics. Data collected by the students will in turn help Vinatzer with his ongoing research.

*Collaboration with Dorothea Tholl.* John Kowalski's biotechnology elective class is participating in a research project with Dorothea Tholl, whose research involves studying the formation of terpenes in the roots of *Arabidopsis* as a chemical defense response against herbivory. An understanding of the role of terpenes in plant defense against herbivores has the potential to open new avenues for designing effective strategies for controlling insect and nematode pests of plants. While working with Tholl, students are examining terpene formation in different ecotypes of *Arabidopsis*. First, the students are growing plants in hydroponics units to enable collection of volatile compounds secreted from roots. Once the plant roots have reached sufficient mass, the students will treat individual plants with jasmonic acid to mimic insect feeding. Students will then use sensitive solid-phase microextraction field collectors to adsorb volatiles emitted by the plant roots. The collectors will be sent to Tholl's lab for analysis using gas chromatography mass spectrometry. The data will then be sent to the students for analysis. Students will have the opportunity to learn a variety of skills including growing plants in hydroponics units, collecting volatile compounds, and analyzing data collected by gas chromatography mass spectrometry, as well as concepts in ecology and bioinformatics. Data collected and analyzed by the students will in turn help Tholl with her ongoing research program.

### 3.3.7   Example of a Student–Teacher–Scientist Partnership

**Living among the Extremophiles: Collaborative Investigations
of Cyanobacteria as a Context for Learning Chemistry and Biology**

*Contributed by Charles K. Jervis, Auburn High School, Riner, Virginia; Judith Jervis, Malcolm Potts, Richard F. Helm, Biochemistry, Virginia Tech.*

#### 3.3.7.1   Introduction

The guiding principle in our partnership has been that honest, productive feedback can lead to an understanding of the partnership's strengths and weaknesses and the sharing of responsibilities in a way that reflects the inherent interests and skill sets of all involved. Given that our objective is for students to develop an appreciation and understanding of science, we work together in a way that actively engages them in the work. In the case of the microbial world, it has been our observation that students like the extreme organisms, the ones that live at "the edge." While the organisms living in the hot springs and hydrothermal vents are fascinating, most of us don't live near such sites and therefore cannot go and observe them as they exist naturally. Thus, we set out to identify an extremophilic ecosystem that is ubiquitous such that students can interact directly with the system, and in doing so, develop a personal understanding of the uniqueness and robustness of the organisms involved. Our ultimate goal is for students to appreciate the microbial community and the scientific processes used to understand them.

### 3.3.7.2   Setting the Stage

The laboratories of Malcolm Potts and Richard Helm are an example of a scientific partnership between a microbial molecular ecologist and an analytical biochemist. While both share an interest in understanding how organisms survive extreme environments, both have very different ways in thinking about the organisms and designing experimental plans. This relationship has spawned a wide variety of projects relating to metabolic arrest and recovery in all types of organisms. One ecosystem that is of particular interest is found on most asphalt shingles roofs in the United States. This immobilized biofilm is dominated by a cyanobacterium presently classified as a species of *Gloeocapsa*. This micro organism withstands temperature extremes (above 100°C in summer and below −30°C in winter), redox active heavy metals (incorporated into the shingles to prevent biofilm growth), erratic cycles of desiccation and rehydration, intense UV-irradiation, and a lack of nutrients as well as hydrocarbons that leach from the asphalt binder. This is an ecosystem generated by the activities of modern civilization, and can be found almost anywhere. Very little is known about the organisms or their mechanisms of survival. Fundamental knowledge concerning metabolic arrest and recovery can be gained from the study of this unique biofilm community, and provides the opportunity for upper-level high school students to participate in the generation of the primary data that are useful for understanding the strategies involved in extremophile community development and maintenance.

### 3.3.7.3   Developing the Relationship

From the outset, we have strived to develop a working relationship that will last over time, mirroring the persistence and tenacity required by those who choose science as a profession. In addition, we wanted students to experience science that didn't always "work," again to highlight that persistence and trial-and-error, as well as experience developed through practice, can yield new knowledge. Although demonstration labs that have predictable outcomes can be used to teach concepts and skills and provide insight into how scientists go about their work, using these lessons to teach about the process and nature of science can be challenging.

We are particularly interested in helping students see the "big picture." How do scientific projects develop? Why do scientists choose some studies over others? How are knowledge bases built up over time? During their undergraduate, graduate, and post-doctoral training, scientists learn the answers to these questions, but most high school students don't have such opportunities. The cyanobacterial ecosystem described above is an ideal context for providing this opportunity in a high school setting. It is ubiquitous poorly understood, and requires little in the way of sophisticated equipment to isolate and grow the organisms. Cooperation between the researchers investigating the system and a high school teacher interested in offering students a first-hand glimpse of the scientific process leads to the development of projects of interest to both.

### 3.3.7.4   Making Connections and Getting Started

In our case, finding the right connection was relatively straightforward. One of us (JJ) has been the lab manager with RFH's research group for over a decade, and is the wife of a high school science teacher (CKJ) who has studied microbial systems in the past. As the lab's cyanobacterial research portfolio grew, conversations about the work led to the development of a teaching and research guide that wraps high school biology and chemistry knowledge and skills around the unifying idea of the chemical biology of extremophile cyanobacteria. Many related concepts and skills were already included in the high school curriculum. The idea was to create an instructional "throughline," an educational concept developed by researchers at Harvard University, to span the academic year by integrating these concepts and skills into the study of cyanobacteria and their applications and implications.

For one summer, CKJ volunteered in RFH's laboratory, developing research skills and a knowledge base for investigating the chemical biology of extremophile cyanobacteria. The following academic year, CKJ piloted the activities in his high school biology and chemistry classes. Students completed them and provided feedback during informal observations and discussions. As a culminating activity, CKJ led his students in a "think and do" field trip (rather than a "watch and see tour" type of field trip) to RFH's laboratory.

To share this work with a broader audience, CKJ and JJ presented the manual and some of its activities during workshops at annual meetings of the Virginia Association of Science Teachers and the National Science Teachers Association. Portions were published in the state association quarterly newsletter. Feedback from workshop participants was also used to modify the manual. In Summer 2006, the manual was used as the primary text in a college level chemistry course for gifted students in the Appalachian Regional Summer Governor's School as part of Virginia's Governor's Schools for the Gifted.

### 3.3.7.5   Making Progress

Students' initial studies involved collecting and documenting samples and extracting secondary metabolites and protein. A summary of this work as well as portions of the teaching manual can be found at: http://vigen.biochem.vt.edu/outreach.

asp. Some students conducted independent research to obtain and analyze secondary metabolite pools by thin layer chromatography. Studies to evaluate the UV-protective and allelopathic properties of these compounds are underway at the high school, with two students working part time in the university research lab. Other students completed directed research on self-selected topics in Helm's lab as part of the research team during the past 2 years.

Developing a mechanism for students to interact with each other and with their teacher and partner scientists is a work in progress. The World Wide Web provides a fast and simple method by which students can share their work, for example, through discussion forums or message boards. In this case, because the students are geographically close, there has been little motivation for them to communicate online. We anticipate that web-based interaction, including discussions about methods and findings, will occur more frequently as the project expands to involve more distant collaborators. Students are also sharing their findings during science fairs, and we expect this continue. In the long term, however, we envision a mini-symposium where the students can get together and discuss their projects in the format of a small scientific meeting.

Are there some who just don't care? How does one deal with those folks? As in any endeavor at this level, and when working with an entire class, there are some who do poor jobs or do nothing to engage the tasks and materials. Some will talk a good talk, but when it comes time to walk the walk, they do not, for various reasons, not the least of which is that sometimes hard work is required. What do we do with these students? To date, students have become involved at all different levels, from just attending class to fully developing an independent research project. We continue to emphasize the opportunities provided during these investigations, giving students the chance to try on different career "hats" and keeping in mind our mantra that successful partnerships can't be forced. Our only advice is just keep trying by giving them opportunities and leading them to the possibilities, bearing in mind that such investigations will be beyond the reach and/or abilities of several in the class. Not everyone will succeed or want a career in science, and we need to remember that they are at the age where they are "trying in different hats" to see which one they may want to wear for a career, and some just don't fit. Nonetheless, it is imperative not let the failures and disconnects of the scientifically-challenged or disinterested hinder their personal growth and development or block the progress of those who can and who chose to try to do something. Relationships cannot be forced.

### 3.3.7.6 Envisioning the Future

As our collaboration continues to develop and mature, we will keep several philosophical and pragmatic questions in mind, including:

- In what ways can we help students understand the value of generating new knowledge?
- How do we identify, adapt, or create demonstration labs that help students develop valuable skills and knowledge that can be applied to discovery and hypothesis testing?

- How do we increase public awareness of neighborhood extremophiles and high-light student contributions to understanding this new avenue of exploration?
- How do we engage more high school students in these kinds of authentic investigations?
- How do we continue to ensure that the teaching manual can be a resource for other teachers and students to use in their study of extremophilic cyanobacteria?
- How can we capitalize on this work to expand the use of multidisciplinary approaches to science learning and scientific investigation?

Finally, how do we build relationships with new partners? We expect that both the students and the science will benefit from collaboration with geographically diverse individuals who have access to ecologically diverse field sites. Lessons learned by answering these questions will allow us to develop this partnership to its optimal extent.

### 3.3.8 Student-Focused Outreach

Certainly curriculum enhancements and student–teacher–scientist partnerships are student-focused. Indeed, most K-12 O&E efforts are geared toward enhancing students' interest in and awareness of science, improving their attitudes toward science and scientists, and ensuring their understanding of science content and process. Student-focused outreach is distinguished here as the group of K-12 O&E activities that involve students yet do not occur in the classroom or with the integral involvement of teachers. For example, there are a number of grassroots and programmatic efforts across the country in which high school students gain science research experience, either in the laboratory or the field (see "International Research Internships for High School Students" in Sect. 3.3.9). Such internships are more likely to succeed if:

- They occur over an extended period of time (e.g., 6–8 weeks of full-time work in the summer, hours-long weekly sessions over an academic year, etc.).
- There is a mentor (e.g., graduate student, post-doctoral fellow, professor) who is willing, interested, and available to supervise the student in their work.
- There is a well-defined project or task that fits the time available for which the student is responsible.
- The materials, supplies, and research context (e.g., field, lab, etc.) are appropriate for pre-college students. The host institution's rules and regulations regarding participation in particular research activities should be reviewed and followed (e.g., Is there a chance the student would be working with or near radioactive materials? What are the rules and regulations regarding minor's work with radiation? Is it appropriate to get parent consent and student assent for the work? In what ways can the student's work with potentially hazardous materials be minimized? If there is field work involved, how will the student get to and from the site? In what ways can the site visit be designed to ensure the student's safety?).

It is also important to keep in mind that students may have little experience with the research endeavor, including how to go about designing and conducting investigations or collecting, recording, and analyzing data. Most professors have some experience mentoring trainees, but early career scientists, such as graduate students and post-doctoral fellows, may need their own mentorship on how to mentor pre-college students. Jo Handelsman and her colleagues at the University of Wisconsin have developed a handbook on mentoring, titled *Entering Mentoring* (2005), which is freely available with support from the Howard Hughes Medical Institute. The advice and exercises included in the guide are useful for mentors and mentees at all levels of experience.

Other models for student-focused outreach can involve larger numbers of students for shorter durations, for example, field trips or summer science camps. Again, joining an existing effort at your institution can help minimize time spent on logistics, including reserving teaching and lab space or arranging meals and lodging. For example, at Virginia Tech alone, there are dozens of science, technology, engineering, and mathematics focused summer activities for pre-college students. C-Tech^2 (Computers and Technology at Virginia Tech; http://www.eng.vt.edu/academics/ctech2.php) is designed to develop and sustain the interests of women in engineering and the sciences, and brings dozens of high school girls onto campus for a two week summer program. C-Tech^2 personnel make all of the logistical arrangements, including recruiting the students, arranging meals, lodging, and social activities, providing resident assistants, and setting the schedule. Interested scientists and engineers are then welcome to "plug into" 1–3 h time slots, during which they can introduce the students to their ongoing research, guide them in lab visits, or lead them in a hands-on activity.

### 3.3.9  Example of Student-Focused Outreach

#### International Research Internships for High School Students

*Contributed by Carla V. Finkielstein, Biological Sciences, Virginia Tech; Daniel G.S. Capelluto, Chemistry, Virginia Tech [additional information available at: http://www.biology.vt.edu/faculty/finkielstein/index.htm; http://www.ottokrause.edu.ar/2k/index2k.htm].*

We originally developed this program as result of a debate that followed a series of lectures organized by the Escuela Nacional de Educacion Tecnica, "Otto Krause," a highly recognized public school in Argentina. The lecture topics focused on scientific and ethical issues that need to be addressed in developing countries. The subsequent debate centered on resolving the scientific needs and technological challenges of developing societies, specifically those in Latin America. Generally in Latin America, students are encouraged to enroll in short technical programs oriented toward the labor market. This may be because policy changes and budget cuts in the public sector are leading to increased pressure on students to acquire knowledge that is more practical in getting a job. Although a larger technical workforce is needed in the short-term, fewer students are pursuing graduate education and research careers, leaving countries unprepared to meet the future needs of an evolving society and the demands of emerging technologies.

Our program aims to address this issue by recruiting and training high school and graduate students in advanced technologies in our area of research while fostering social interaction with local students and promoting collaborations with their international institutions. We anticipate that these social interactions, both locally and globally, will encourage students to pursue careers in science research as well as initiate international collaborations that will be fruitful for countries at all stages of economic development. Thus, the project's overarching goal is to enhance global science education while maintaining a commitment to cultural diversity, cross-cultural tolerance, professional ethics, and international skill development. We hope to achieve this by providing authentic science research experiences to high school students with diverse cultural and socio-economic backgrounds.

Argentinian students are recruited from the Escuela Nacional de Educacion Tecnica, "Otto Krause," which emphasizes technological areas, including engineering, computer sciences, chemistry, and mechanics. Otto Krause is among the top public schools in Buenos Aires, with a female enrollment of only 7%. Upon completion of a 6-year program, students are encouraged to continue their education in areas of applied sciences. Finkielstein, one of the principal investigators of this program, is a former student in the chemistry track and, thus, is intimately familiar with the educational resources of the school and the quality of its education. Otto Krause students are among the best-prepared professionals in the country, yet lack of funding has limited students' access to emerging technologies. In the U.S., many students have opportunities to learn about these technologies, but have a less international perspective on their education. Our program aims to fill this gap by bringing U.S. and Argentinian students together, promoting their academic development while encouraging them to become familiar with each other's culture.

To date, we have focused on putting the international components of the program into place. Positions are advertised at Otto Krause by professors, in the students' monthly newsletter "Contactos," and on Finkielstein's web site. Students apply during their senior year of high school to be eligible for the following year internship. The application process involves several phases: (1) Every applicant is assigned a number to avoid bias in the selection process, (2) Students complete a questionnaire in both Spanish and English, stating their career goals, commitment to higher education, community service experience, other interests, and grades; (3) Students provide reviewers with two letter of recommendation, one from a current and/or former professor and one from a friend, who must address non-academic issues. The top five candidates are invited for interviews. Selected students are notified seven months in advance to facilitate travel and visa arrangements.

At least three months before their arrival, international students receive a "Research Package" with an outline of the experimental work to which they will contribute over the next year. Students participate in intra-lab rotations where they learn the basics of molecular and cellular biology while generating reagents that are needed for various projects in the lab. Rotations include five areas of interest and are designed for the student to learn cutting-edge technologies: (1) DNA manipulation,

Expression and Purification of Recombinant Proteins, (2) *In vivo* model systems, (3) Biophysical Approaches – Functional Assays, (4) Microscopy, and (5) Bioinformatic Analysis of Genome Sequences. Experiments are planned week-to-week and summarized at the end of each rotation. Each student has a specific and unique project and works along with a graduate student or postdoctoral scientist in the laboratory, who is the primary mentor in that area of research. The research plan is complemented with social events and weekend trips to surrounding areas. The students stay in dorms and are encouraged to participate in on-campus life. This gives them the opportunity to experience a different educational and geographical setting, consider college education, learn and appreciate new cultures, and form lasting friendships.

Virginia Tech and its Department of Biological Sciences host the program. The university facilitates the acquisition of visas for participating students and provides on-campus housing. Research activities occur in Finkielstein's lab in the Department of Biological Sciences in collaboration with Capelluto in the Department of Chemistry. Funding for international students' training comes from a number of sources. The Escuela "Otto Krause" supports some travel and day-to-day expenses. The Department of Biological Sciences covers students' visa expenses. Finkielstein's intramural funding pays for housing and experimental materials and supplies. Finkielstein is also seeking funding from extramural sources, including the National Science Foundation, to support this initiative.

This program has been in place for the past two years. Students have demonstrated their progress in a number of ways. First, they present a seminar at the end of their stay attended by all members of the laboratory and their collaborators. Second, students summarize their experience in a series of public lectures organized by their home schools. Finally, students are choosing to pursue undergraduate education in science rather than moving directly into the workforce. For example, two previous participants graduated last fall and enrolled for tertiary education in the School of Natural Sciences, University of Buenos Aires, majoring in Molecular Biology and Chemistry. Students attending this year's program will graduate in Fall 2007 and have already enrolled in tertiary education at the University of Buenos Aires, Argentina.

## 3.3.10  Professional Development

The National Research Council has authored a guide on scientists' involvement in teacher professional development (1996), which includes a large appendix of example programs. Also included is information about program structure and the variety of roles for scientists as well as advice for how to initiate, implement, evaluate, and fund the effort. Thus, only a few highlights are mentioned here. Teacher professional development programs are structured in many ways, from few hour-long workshops to yearlong intensive programs. Topics can range from learning about current research (see "Current Science Seminar Series" in Sect. 3.3.11) to developing greater familiarity with the science education standards to learning a new laboratory or teaching technique. Professional development efforts are more likely to be useful and appealing to teachers if they:

- Are planned to fit teachers' schedules (e.g., are held in the summer but not right before the school year starts or right after it ends).
- Are coordinated with professional development regularly planned by schools and districts (e.g., occurring during a regularly scheduled professional development day, during which school is not in session or substitutes are available to teach).
- Offer graduate credit or continuing education units that can help teachers make progress toward recertification (i.e., in most states, teachers must participate in a minimum number of hours of professional development to maintain their licensure).
- Provide a concrete way to translate what is learned through professional development to classroom practice (e.g., time to create or adapt lessons, materials and supplies for classroom use, etc.; see "Wyoming Student Teacher Education Program (WySTEP): Learning to Teach Science Through Hands-on Research" in Sect. 3.3.12).
- Compensate teachers appropriately (i.e., if teachers are committing their professional time, for example in adapting lessons for classroom use or participating in research, they should be paid accordingly).

Programs through which scientists grow professionally are becoming more common. For example, NSF's Graduate Research Fellows in K-12 Education (GK-12; http://www.nsf.gov/funding/pgm_summ.jsp?pims_id=5472&org=DGE) is designed explicitly to encourage professional growth of participating graduate students:

> This program provides funding to graduate students in NSF-supported science, technology, engineering, and mathematics (STEM) disciplines to acquire additional skills that will broadly prepare them for professional and scientific careers in the 21st century. Through interactions with teachers and students in K-12 schools and with other graduate fellows and faculty from STEM disciplines, graduate students can improve communication, teaching, collaboration, and team building skills while enriching STEM learning and instruction in K-12 schools. Through this experience, graduate students can gain a deeper understanding of their own STEM research. In addition, the GK-12 program provides institutions of higher education with an opportunity to make a permanent change in their graduate programs by incorporating GK-12 like activities in the training of their STEM graduate students.

### 3.3.11 Example of Complementary Professional Development for Teachers and Scientists

**Current Science Seminar Series**

*Contributed by Katherine Nielsen, Rebecca Smith, Andrew Grillo-Hill, Patricia Caldera, Chantell Johnson, and Laura Gibson, Science and Health Education Partnership, University of California at San Francisco.*

Initiated in 1987 by University of California at San Francisco (UCSF) professors Bruce Alberts and David Ramsey, the Science and Health Education Partnership (SEP) is recognized nationally and internationally as a model organization that promotes partnership between scientists and educators in support of high quality science education for K-12 students (http://biochemistry.ucsf.edu/%7Esep/index. html). To this end, SEP's programs: (1) support teaching and learning among teachers, students, and scientists; (2) promote an understanding of science as a creative discipline, a process, and a body of integrated concepts; (3) contribute to a deeper understanding of partnership; and (4) provide models and strategies for other institutions interested in fostering partnerships between scientific and education communities.

In 2006–2007, over 250 volunteers contributed more than 20 h each in San Francisco Unified School District (SFUSD) K-12 classrooms. These volunteers include graduate students, postdoctoral fellows, researchers, and professional students (medical, pharmacy, dentistry, nursing, and physical therapy). On a yearly basis, SEP works with 85–90% of the 120 public schools in San Francisco and approximately 300 teachers. SEP offers a diverse programmatic menu: a variety of classroom-based partnership models which bring UCSF volunteers into K-12 classrooms, summer courses and seminars for teachers, professional learning communities focused on science teaching, and a high school internship program. Additionally, teachers and UCSF volunteers borrow materials from SEP's lending library and these materials provide enriched science learning experiences for over 20% of SFUSD's students.

One of SEP's newer programs, the *Current Science Seminar Series* (*CSSS*), brings middle and high school science teachers to UCSF for interactive seminars on current research in biology led by UCSF early-career scientists. SEP Coordinators coach the scientists in the development of their presentations through both a workshop introducing interactive presentations and individual practice sessions. Presentations are designed to be accessible to middle and high school science teachers and all include at least one interactive component. Teacher participants give constructive oral and written feedback to scientist presenters to help them make improvements for future presentations.

In 2006–2007, *CSSS* engaged 20 teachers[1] and five scientists in investigations and conversations around science, provided a forum for discussion of grades 6–12 science curriculum development, and facilitated networking among teachers from schools across the district. *CSSS* was hosted from 4:30 to 6:30 pm and dinner was provided. The five scientists included three postdoctoral fellows, one professor, and one graduate student. Middle and high school science teachers were invited to the seminars through their SFUSD Science Department Chairs and existing SEP teacher network contacts. Teachers participating in four or more of the seminars received credit through University of California at Berkeley Extension.

SEP Coordinators met with the scientists once before their talks to assist them in designing their presentations and in the inclusion of hands-on components. Activities in the five seminars differed based upon the knowledge of the presenter, their session content, and access to resources. Activities ranged from ones that were

---

[1] There are a total of 107 SFUSD middle and high school life science teachers, so attendance represents approximately 20%.

materials intensive (e.g., examining histology slides of breast cancer biopsies or using Lego models to demonstrate a genetic screen) to others that required little or no materials (e.g., think-pair-share or small group discussions). Four out of five seminars included a visit to the presenters' research laboratory to enhance the learning experience of the teachers that attended. All of the seminars were videotaped and made available to all teachers through SEP's lending library. Teachers were able to watch videos of the seminars they could not attend or revisit seminars they found of interest. Also, the speakers used their own videos to learn more about their presentation style. These were the seminars presented in the spring of 2007:

- Viruses, Genechips, and Disease Using Technology to Advance Understanding of Viral Disease
- Parasites and You: Signaling Pathways in Schistosomes
- Drunk Worms: *C. elegans* and Alcohol Addiction
- Imaging the Dynamic Brain: Learning About How We Learn
- Imaging Cancer Under the Microscope: Understanding the Pathology of Breast Cancer

Scientist presenters reported that this experience increased their appreciation for active teaching methods. They may have had prior exposure to such methods, but having an opportunity to try them out reinforced their effectiveness. They especially liked using interactive techniques (e.g., think-pair-share). One presenter noted:

> I will try to incorporate different active learning techniques. Before this series, it was not obvious to me whether this helps the audience, but now from my experience with CSSS, I could see for myself that it does. The typical research presentation that I participate in does not encourage this type of teaching/presentation.

Learning from scientists who are currently working in the field was exciting for teachers and many expressed that this was something that they had been missing. They were inspired by the research presented.

- 100% strongly agreed/agreed that this series has sparked their interest in learning more about one or more of the seminar topics.
- 100% strongly agreed/agreed that the series conveyed the dynamic nature of scientific knowledge or understanding.
- 73% strongly agreed/agreed that in the future they plan to read scientific or medical news/magazines/journals more than they do now.

As one teacher wrote: I gained a broader appreciation for the different types of research that UCSF does.

Overall, there are a variety of reasons that teachers attend *CSSS*. Some miss the connection they used to have to cutting-edge scientific research and just want to learn new things. Others have personal interest in a specific topic and wish to know more. Also, at every seminar, there was at least one participant who came looking for ways to incorporate that particular content into their classroom. They were sometimes already doing a unit on the topic or already using similar methods (e.g., using worms in the classroom), and other times wanted to use this content for a new lesson.

### 3.3.11.1   Future Directions

As we begin the second year of *CSSS*, there has been great interest in the program from early career scientists. Twenty scientists at UCSF applied for the seven available seminars in the 2007–2008 series. This year we have also for the first time invited high school teachers to bring one or two of their students with them; at our first seminar (November 2007), the students were thrilled to be able to attend. This program will continue through 2010 with funding from a Howard Hughes Medical Institute Undergraduate Science Education Award.

### 3.3.11.2   One Scientist's Experience

Jon Pierce-Shimomura, one of the spring 2007 presenters, shared the following about his *CSSS* experience:

Due to the lengthy time course of a research career, it had been about ten years since I had last taught a class. As such, I was eager to participate in *CSSS* to refresh my teaching experience before pursuing a teaching position. I was also excited to share my research results with junior high and high school teachers who could transfer this knowledge to hundreds of students. I found that the step-by-step assistance in the design of our seminars was very useful. SEP Coordinators first moderated a forum where participating researchers discussed methods for effective teaching, topics that would be of interest to our audience and ways to enliven our seminars with interactive components. For instance, after considering the "less is more" strategy, I narrowed down my seminar from "Mechanisms of drugs and alcohol on the nervous system" to only cover "alcohol." I was also encouraged to include two interactive components: one that could be completed on site (we used fluorescent microscopes to view color-coded neurons of living worms) and another that could be adapted in their school classrooms (we used LEGO models of molecules to demonstrate the concept of a genetic screen for targets of drugs and alcohol).

SEP Coordinators next set up practice sessions for our seminars. Here I received helpful input in adjusting the content for a non-researcher audience with supportive diagrams and movies. SEP Coordinators provided videotaping of our final seminar that I have used to think of ways to improve my style of elocution. Overall, my seminar appeared to succeed based on written feedback from the teachers. Lastly, my experience with *CSSS* has fostered my relationship with one of the participating high school teachers. Together we designed a portable curriculum for "How to use model systems to study the effects of drugs and alcohol on the nervous system." The curriculum includes lesson plans, worksheets, video presentation and an affordable interactive component – quantitative behavioral tracking of worms with video microscopy (less than $500 set up costs). I fully expect that the *CSSS* program will continue to provide valuable experience both to researchers in teaching effectively and to teachers in knowledge on cutting edge research.

## 3.3.12 Example of Teacher Professional Development

*Wyoming Student Teacher Education Program (WySTEP):*
*Learning to Teach Science Through Hands-on Research*

*Contributed by Anne W. Sylvester, Department of Molecular Biology, University of Wyoming (annesyl@uwyo.edu); Joseph Stepans, College of Education, University of Wyoming [additional information available at: http://epscor-wise.uwyo.edu/ WySTEP/WySTEP_Information.htm].*

### 3.3.12.1 Description of the Project

Improving science literacy on a national scale requires re-evaluating all levels of teaching and learning in the United States. The University of Wyoming (UW) has addressed this national challenge by initiating a new science education program targeted for pre-service secondary education majors. The Wyoming Student Education Program (WySTEP) is designed to introduce future science teachers to hands-on scientific research, thereby preparing the students to integrate research and basic scientific inquiry into the secondary school curriculum. WySTEP is a collaborative effort among all sectors of education including administrators and mentor teachers in Wyoming secondary schools, UW College of Education faculty, UW science faculty, graduate students, and the participating science education majors. The program is developed in response to a need in the state and nation to find more effective ways for future science teachers to understand the processes and nature of science. In its second year of implementation, the pilot program is currently funded by the National Science Foundation's Experimental Program to Stimulate Competitive Research (NSF EPSCoR), an NSF award to Wyoming.

### 3.3.12.2 Underlying Goals of WySTEP

Children are natural scientists because they have inherent curiosity and driving interest in the world around them. By the time these natural scientists reach secondary school age, that drive is often eliminated, rechanneled, or buried in the secondary school system. We believe the spark of scientific interest can be recaptured in secondary school students if science is taught in its full diversity as an observational, often hypothesis-driven, discovery based, often collaborative discipline that opens doors of analytic and creative thinking for students. The most systematic way to reach these students is through their teachers, who plan year-long science curricula and day-to-day science experiences for their students. Yet the very students who will be future teachers are also the former students who missed the opportunity to learn about the true nature of science. To break this cycle, WySTEP is designed to provide secondary science ed majors with a summer research experience under the guidance of doctoral students in science disciplines and the research is then translated into a tangible classroom experience. Development of the unit is key to connecting research to teaching and maximizes the value of the research experience for future teachers.

WySTEP aims to accomplish several goals. First, the science ed major will gain a deeper understanding of the research process and will develop self-confidence in understanding hypothesis testing and experimental design. Secondly, the student will learn how to translate hands-on research into inquiry-based exercises for the secondary classroom. This process is highly collaborative because it engages the secondary school mentor teacher, who supervises student teaching and provides direct classroom guidance, the mentoring graduate student, who guides the research process, and education faculty, who facilitate translation of the research into an inquiry unit for classroom use. Finally, the program aims to provide mutual benefit to all participants by creating an environment of educational exchange. For example, while mentoring, graduate students also learn about the value and challenges of teaching from the science ed majors. Similarly, while overseeing the execution of the lab unit, the high school mentor teachers gain or renew their appreciation for the research enterprise.

### 3.3.12.3  Specifics of WySTEP

The year-long program begins for the secondary science ed major during the junior year, one year prior to student teaching, and ends the next spring, after student teaching. As detailed below, each semester of the WySTEP cycle involves specific activities, guidance, and collaboration among participants.

*Spring:* To begin the program, ed majors are selected through a competitive application process to be WySTEP Fellows, and are then paired with a graduate student. The Fellow and the graduate student both receive stipends for participation. If secondary school mentor teachers have been assigned, all three parties meet to discuss directions, needs, and plans. Mentor teacher buy-in is essential at the outset and is required for development of a new classroom unit.

*Summer:* The graduate student serves as a research mentor to the WySTEP Fellow during the summer research experience. UW faculty meet with the pair periodically to discuss progress and to facilitate the conceptual translation of research to the development of a teaching unit.

*Fall:* WySTEP Fellows formulate an inquiry based classroom unit through consultation with education faculty, the mentor graduate student, and the mentor secondary school teacher. The unit is presented to the entire group prior to student teaching.

*Spring:* WySTEP Fellows disperse throughout Wyoming for student teaching assignments, where the unit is implemented in the classroom. Graduate students travel to the teaching site if needed to assist with execution of the unit. WySTEP provides permanent research equipment and supplies to the mentor teacher and school, if needed. WySTEP Fellows convene to give a presentation at Undergraduate

Research Day (http://epscor-wise.uwyo.edu/Research_Day_2007/researchday_2007. htm), closing the program.

#### 3.3.12.4  Evaluation of Program Impact

WySTEP is currently in its second year, with the first class of Fellows graduated in spring 2007. Of these students, 83% were immediately employed in teaching posts and 17% went on to graduate school in science education ($n = 6$). Methods are being developed to track short- and long-term outcomes. Anecdotal reports of impacts accumulate rapidly in a state like Wyoming, where the population of 500,000 is surprisingly interconnected. Specific data for quantitative assessment in the future will include tracking graduates of the program by periodic questionnaires.

#### 3.3.12.5  Program Dissemination, Challenges and Development

The first year of WySTEP proved highly successful based on feedback from students and mentor teachers. Program challenges include difficulties in coordinating all participants. A long-term goal, not yet achieved, is to include high school administrators so that program values can be incorporated into school administrative planning. Currently, limited funds provide stipends to six pairs of WySTEP Fellows and graduate students per year, thus limiting the program to local students only. We would like to expand the opportunity, provide longer research experiences, advertise more widely, and facilitate statewide acceptance so that a permanent funding stream can be established. We hope that by working directly with our future teachers, we can influence the science literacy of students for generations to come.

### 3.3.13  Other Venues

Although the projects described and the strategies suggested here are the result of efforts with K-12 students and teachers, many of the lessons learned through K-12 O&E are applicable to other venues. For example, there are many non-formal programs, such as 4-H (http://www.4-h.org/), Future Farmers of America (http://www.ffa.org/), Girl Scouts of the U.S.A. (http://www.girlscouts.org/), the Boys and Girls Clubs of America (http://www.bgca.org/), and many other organizations, that involve educators or other adult mentors in working with youth. The same principles apply: identify needs, match needs and resources, take advantage of existing efforts, and collaborate with individuals who know the learners (e.g., 4-H agents, youth group leaders, etc.). These non-formal venues can offer a level of flexibility not available in schools because standards regarding science learning are not typically applicable. Yet, some programs have curricula that are worth exploring in the same ways as the standards, to determine age appropriateness, minimize "reinventing the wheel," and integrate into the flow of the program to maximize the potential for impact.

Non-formal learning is distinguished from informal learning (e.g., science centers, museums, aquaria, etc.), because non-formal programs are often somewhat structured (e.g., youth visit repeatedly on a regular schedule, activities can occur over several visits, activities are mentored or facilitated, etc.). Informal venues can be home to non-formal programming, but largely have a voluntary audience who visits only one or a few times or on an unpredictable schedule (i.e., not daily or weekly like in schools or non-formal programming). The participants' engagement with more knowledgeable or experienced individuals, like docents, is also voluntary, and materials and exhibits are usually designed to "stand alone."

One advantage of working in informal venues is the opportunity to reach a much larger audience. Exhibitions at science centers in metropolitan areas may have millions of visitors. In addition, some exhibitions are portable or have movable components and can be made available for viewing and interaction by people across the country through relationships among science centers. For example, Plants-in-Motion developed by plant biologist Roger Hangarter and colleagues at Indiana University is a web-accessible series of time-lapse videos that can be used to illustrate concepts in plant development and physiology. Hangarter collaborated with artist Dennis DeHart at Buffalo State University of New York to develop the sLowLife exhibition, which has a virtual component available via the web (http://plantsinmotion.bio.indiana.edu/usbg/) and an in-person exhibit that is rotating among science centers and botanical gardens across the country (http://www.chicagobotanic.org/slowlife/index.html).

This example demonstrates another advantage of informal learning, the flexibility available with respect to medium and content: educational materials do not have to fit into a prescribed curriculum or align to a specific set of standards. Informal venues tend to attract self-selecting audiences who are already enthusiastic or knowledgeable about science. If your goal is to engage a broader audience of learners who may not choose to seek science education opportunities in their free time, formal settings (i.e., classrooms and schools) may be your best bet. Also, just as work in K-12 schools is best done in collaboration with teachers or other school personnel, collaborations with informal educators can help to maximize the use and usefulness of informal science learning efforts.

### 3.3.14 Additional Points to Consider

Most importantly, make a finite commitment with defined expectations (*Moreno 2005*). Keep in mind that many of the constraints K-12 students and teachers experience in their classrooms are magnified versions of those we experience in teaching our own courses (McKeown 2003). Budgets are limited, time is tight, space is inadequate, and there is too much content to "cover." Teachers have the added challenges of addressing standards mandated at the national, state, district, and even school levels, being held responsible for this work based on their students' test scores even to the point of being paid based on student achievement. Teachers have little to no time during the day to plan lessons, prepare lab materials, grade student work, respond to student and parent concerns, or even eat lunch, thus much of the collaboration may need to occur after their school day has ended. Think through the logistics of your plan, as well as how you will communicate about and alter the plan if and when contingencies arise. What do you plan to do? When will activities start and end? Whose responsibility is it to gather materials, lead instruction, and carry out the evaluation? Ultimately, as experienced teachers know, not everything can be planned or anticipated and you will have to "just do it" (Lally et al. 2007).

## 3.4   Funding the Effort

Excitingly, because of widespread interest in public education, especially related to science and technology literacy, funding is available at the local, state, and national level. Personnel whose experience and expertise involves fund-raising or development can be strong partners in the search for support. As when seeking any extramural funds, it is important to consider the interests of those who want to invest in science teaching and learning. Local business, for example, local industry, banking, and franchises, as well as benefactors who live in the community are often willing to provide smaller grants of several hundred to a few thousand dollars. Larger scale funding is available from federal agencies, including the National Institutes of Health, the National Science Foundation (NSF), the Department of Education, and other entities. Major grants are also available from private foundations and other charitable organizations (Table 3.2). As noted in Chap. 1, NSF and other federal agencies are finding creative ways to encourage scientists to become involved in K-12 education through supplemental funds to current grantees (http://www.nsf. gov/bio/supp.jsp), including monies to:

- Provide Research Experiences for Teachers (RETs; NSF-05-524)
- Provide Research Assistantships for Minority High School Students (RAMHSS; NSF-89-39)
- Generally broaden the impact of research, including involving K-12 audiences, as part of science grant activities ("Examples of Broader Impacts" at: http:// www.nsf.gov/dir/index.jsp?org=BIO)
- Communicate Research to Public Audiences (http://www.nsf.gov/pubs/2003/ nsf03509/nsf03509.html; Note: These funds are not available for K-12 classroom activities, but may involve students and teachers in other contexts, for example, through after-school or museum-based programming)

If you are seeking funding at the national level, consider how your efforts can have national appeal. For example, consider incorporating a plan to disseminate lessons learned, program outcomes, or other products at a national level (see Chap. 5).

Alternatively consider how your efforts serve as a nationally applicable model. Finally, take advantage of existing resources in the form of investment by your own institution. Consider dedicating part of a person's time, including your own, even when you are seeking extramural funds. This can be a powerful demonstration of your commitment to the effort, which can be leveraged to garner additional support.

# References

Bybee R, Morrow C (1998) Improving science education: The role of scientists. Newsl Forum Educ Amer Phys Soc. http://www.spacescience.org/Education/ResourcesForScientists/Workshops/Four-Day/Resources/Articles/1.html. Accessed 3 December 2007.

Cutler S, McCourt P (2005) Dude, where's my phenotype? Dealing with redundancy in signaling networks. Plant Physiol 138:558–559.

Dolan EL, Soots BE, Lemaux PG, Rhee SY, Reiser L (2004) Strategies for avoiding reinventing the precollege education and outreach wheel. Genet 166:1601–1609.

Elgin SCR, Flowers S, May V (2005) Modern genetics for all students: An example of a high school/university partnership. Cell Biol Educ 4:32–34.

Evans CA, Abrams ED, Rock BN, Spencer SL (2001) Student/scientist partnerships: A teachers' guide to evaluating the critical components. Am Biol Teach 63:318–323.

Frechtling J (2002) The 2002 user friendly handbook for project evaluation. National Science Foundation, Arlington VA. http://www.nsf.gov/pubs/2002/nsf02057/. Accessed 25 April 2007.

Handelsman J, Pfund C, Lauffer SM, Pribbenow C (2005) Entering mentoring: A seminar to train a new generation of scientists. University of Wisconsin Press, Madison WI. http://www.hhmi.org/catalog/main?action=product&itemId=272. Accessed 3 October 2005.

Lally D, Brooks E, Tax FE, Dolan EL (2007) Sowing the seeds of dialogue: Public engagement through plant science. Plant Cell 19:2311–2319.

McKeown R (2003) Working with K-12 schools: Insights for scientists. BioScience 53:870–875.

Moreno N (2005) Science education partnerships: Being realistic about meeting expectations. Cell Biol Educ 4:30–32.

National Research Council. (1996) The role of scientists in the professional development of science teachers. National Academies Press, Washington DC.

Oberhauser KS, Brinda SJ, Weaver S, Moon RD, Manweiler SA, Read N (2006) Growth and survival of monarch butterflies (Lepidoptera: Danaidae) after exposure to permethrin barrier treatments. Environ Ent 35:1626–1634.

Solensky MJ, Larkin E (2003) Temperature-induced variation in larval coloration in Danaus plexipppus (Lepidoptera: Nymphalidae). Ann Ent Soc Amer 96:211–216.

Tomanek D (2005) Building successful partnerships between K–12 and universities. Cell Biol Educ 4:28–29.

# Chapter 4
# How Well?

## Evaluating K-12 Outreach and Engagement

Regardless of the path of activities you choose to pursue in K-12 outreach and engagement, evaluation will help you know when it "works" and help you learn when it doesn't. In this chapter, evaluation is defined and distinguished from research and assessment. Ways to approach evaluation are described, including the use of qualitative and quantitative methods and consideration of evaluation findings in formative and summative ways. Also included are a variety of resources useful for guiding the design and implementation of evaluation plans.

## 4.1 Introduction

Although it is not within the scope of this text to fully describe the methods and nuances of science education evaluation and research, it is essential to ask: how will you know that what you are doing works? The first step is to determine what evidence will convince you and your school colleagues that you are achieving your intended goals. Start by discussing what data would be evidence that you are achieving your goals, and how you could go about collecting and analyzing these data and reflecting on the results. Data can be collected more or less formally in a number of ways, including both qualitative and quantitative methodologies (Anfara et al. 2002; Ercikan and Roth 2006; Frechtling 2002; Stufflebeam 2001; Sundberg 2002). If you are interested in interpreting the outcomes and impacts of your effort in a more systematic, rigorous, and generalizable way, consider collaborating with an education evaluator or researcher, or a graduate student in education who is mentored by faculty with appropriate expertise and is seeking a dissertation project.

Because K-12 O&E activities come in many shapes and sizes, strive to achieve myriad goals, and have varying access to resources and expertise, there is no single approach to designing or conducting an evaluation. Evaluations are equally diverse in their scope, as well as in the regulations and resources that influence their planning and implementation. Yet, there are commonalities that provide insight into the what, when, where, who, and how of evaluation. Most importantly, the guiding question that propels all evaluation is: "What do you want to know and how will you know it?"

E.L. Dolan *Education Outreach and Public Engagement*,
DOI: 10.1007/978-0-387-77792-4_4 © Springer Science+Business Media, LLC 2008

## 4.2   Definitions: What is Evaluation?

As outlined in the Joint Committee Standards for Educational Evaluation (1994), the purpose of evaluation is to determine the value of something: its merit (i.e., excellence), its worth (i.e., its excellence and utility, which differs from merit because an activity may be of high quality, but too expensive to sustain), its probity (i.e., adherence to moral standards), or significance (i.e., its impact or importance). In addition, quality evaluation must adhere to four principles: utility (are the findings useable by intended users?), feasibility (is it viable?), propriety (is it legal and ethical?), and accuracy (are the findings valid?).

On the contrary, research is intended to add to a body of knowledge. Evaluation findings are not always appropriate for sharing in a scholarly venue, and research does not always have the immediate usefulness of evaluation. That said, research and evaluation can be conducted simultaneously, but this must be reflected in the evaluation design. Data collection for research and evaluation must be considered explicitly in the goals for both the program and its evaluation. Although there is some controversy about the distinction between "evaluation" and "assessment," the generally accepted understanding is that assessment is the measurement or characterization of something or someone, while evaluation is an appraisal. In other words, assessment is the process of documenting knowledge, skills, attitudes, and beliefs, often in measurable terms. In contrast, evaluation involves systematically determining merit, worth, probity, or significance based on the results of one or more assessments.

Three questions frame any successful evaluation design: what do you want to know, who are your stakeholders and what do they want to know, and what resources are available to conduct the evaluation? The role that each of these questions plays depends on the current and future goals of the evaluation. Evaluation may be intended to determine the needs of individuals participating in the partnership (i.e., needs assessment) to inform the design and ongoing implementation of the partnership (i.e., formative assessment). Alternatively, the individuals involved may be interested finding funding to support, expand, or replicate the program. Different funders are interested in supporting programs with different anticipated outcomes (e.g., student learning, student career choice, enhanced opportunities for under-served students), and a compelling evaluation should be designed with the intent of collecting data related to those outcomes.

For example, if the goal of a series of K-12 O&E activities is to enhance elementary students' attitudes about and understanding of science, possible student outcomes include: (1) changes in their perceptions of scientists as depicted in their drawings of scientists before and after the activities, (2) improved scores on standardized tests, (3) changes in their enthusiasm about science as revealed through focus group discussions or conversations with their parents, (4) increased enrollment in more challenging middle school science courses, (5) increased enrollment in high school science courses, (6) increased pursuit of an undergraduate science major, or (7) increased pursuit of careers in science. Some of these outcomes will be straightforward to document, while others are likely to be beyond the scope, interest, or available time and resources of the program.

## 4.3　What do You Want to Know?

A good starting point for defining the scope of the evaluation is to ask: who has an interest in the results and how will the results be used? In other words, who are the stakeholders and what will they do with the evaluation findings? If the individuals who are involved in the day-to-day K-12 O&E activities are the primary stakeholders, for example a collaborating scientist and teacher, then the findings may be used solely to make improvements for the next time they work together (i.e., formative evaluation). If they are interested in documenting what the students learned, the teacher may want to consider the evaluation results in deciding students' grades and the scientist may share the results in grant proposals seeking additional support for the activities (i.e., summative evaluation).

Revisiting the logic model, described in Chap. 3, can be helpful for identifying predicted outcomes and determining ways to document them. An evaluation matrix (Table 4.1) can be developed by expanding the outcomes component of the logic model to include: evaluation questions, data collection approaches or sources of evidence, respondents (i.e., from whom will the data be collected), and schedules (i.e., when is most appropriate to collect the data). Evaluation questions are usually

**Table 4.1**　Sample evaluation matrix

| Question 1. What do students learn about XYZ as a project participation? | | | |
|---|---|---|---|
| Sub-question | Schedule | Data collection approach | Respondents |
| 1a. What do students learn about XZY concepts and methods? | Course start/end | Pre/post test | Students |
| | Ongoing | Survey | Students |
| | Course end | Instructor phone interviews | Instructors |
| 1b. What do students learn about how the relationships among XYZ concepts and methods? | Course start/end | Pre/post test | Students |
| | Ongoing | Survey | Students |
| | Course end | Instructor phone interviews | Instructors |
| 1c. What do students learn about the relevance of XYZ concepts and methods? | Course start/end | Pre/post test | Students |
| | Ongoing | Survey | Students |
| | Course end | Instructor phone interviews | Instructors |

| Question 2. Is student understanding of scientific inquiry expanded as a result of project participation? | | | |
|---|---|---|---|
| Sub-question | Schedule | Data collection approach | Respondents |
| 2a. What do students learn about scientific inquiry and experimental design? | Course start/end | Pre/post test | Students |
| | Course end | Poster presentations | Students |
| | Ongoing | Survey | Students |
| | Course end | Instructor notes | NA |
| | Course end | Instructor phone interviews | Instructors |

(continued)

**Table 4.1** (continued)

*Question 3. Is the proposed activities were created, piloted, and disseminated successfully?*

| Sub-question | Data collection approach |
| --- | --- |
| 3a. What evidence suggests that the product/project was created? | Educational materials<br>Revised educational materials |
| 3b. What evidence suggests that the product/project was piloted? | Course catalog<br>Syllabus<br>Enrollment records |
| 3c. What evidence suggests that the product/project was disseminated? | Conference presentations and participant list<br>Implementation by participants |
| 3d. What evidence suggests that the evaluation results are disseminated? | Conference presentation<br>Manuscript submission |

the restatement of goals or objectives in question format. For example, if an objective is for students to learn about concepts in classical and molecular genetics, an appropriate evaluation question would be: what do students learn about concepts in classical and molecular genetics? More specific sub-questions can then be developed. For example:

- What do students learn about the role of genes in determining traits?
- What do students learn about the role of the environment in regulating gene expression?
- What do students learn about the inheritance of genes? etc.

Once sub-questions have been identified, sources of evidence can be considered. For example, teachers and scientists may be able to get a sense of what students learn about these concepts by talking informally with students during class, reviewing the students' work, discussing the topics directly with groups of students, and examining changes in students' responses on a pre/post test of genetics knowledge. The corresponding data sources would be: transcripts or notes from interviews with teachers and scientists, students' work, transcripts or notes from focus groups with students, videotapes or field notes from classroom observations, and pre/post test results. As a specific example, Table 4.2 outlines the possible sources of evidence that the goals of the Partnership for Research and Education in Plants (PREP) are achieved. The following resources provide detailed guidance on developing evaluation plans:

- W. K. Kellogg Foundation web site under Knowledgebase, Toolkits, Evaluation (http://www.wkkf.org/, accessed 11/30/2007)
- Program Development and Evaluation through the University of Wisconsin Cooperative Extension (http://www.uwex.edu/ces/pdande, accessed 11/30/2007)
- NSF User-Friendly Handbook for Mixed-Methods Evaluation (Frechtling 2002)

**Table 4.2** How will you know? Possible sources of evidence that Partnership for Research and Education in Plants (PREP) goals are achieved

| Program goals | Sources of evidence that goals are achieved |
| --- | --- |
| **For students** | |
| Learn about plants, genetics, and genomics in the context of doing original research | • Student lab reports<br>• Student presentations<br>• Other student work<br>• Pre/post tests |
| Become more proficient in conducting scientific investigations | • Student lab reports<br>• Student presentations<br>• Student focus groups |
| Contribute original data to current scientific investigations | • Scientist interviews<br>• Scientific presentations or manuscripts acknowledging students' contributions |
| Interact with scientists | • Classroom visits by scientists<br>• Distant interactions (e.g., email, video chat, phone calls, etc.)<br>• Student focus groups<br>• Teacher and scientist interviews |
| **For teachers** | |
| Update plant biology, genetics, and genomics aspects of curriculum | • Teacher interviews<br>• Documents (e.g., syllabi, curricula, homework assignments, etc.) |
| Provide an authentic research experience for students | • Teacher interviews<br>• Student focus groups<br>• Scientific presentations or manuscripts acknowledging students' contributions |
| Gain insights in how to manage student research that does not have a predictable outcome | • Scientist interviews<br>• Conference presentations or manuscripts acknowledging students' contributions |
| **For scientists** | |
| Enhance science learning of students who will soon be in university classrooms and labs | • Teacher interviews<br>• Student focus groups |
| Gain scientific insights from high school students' work | • Scientist interviews<br>• Scientific presentations or manuscripts acknowledging students' contributions |
| Broaden the impact of research by expanding the boundaries of research lab into the community | • Scientist interviews<br>• Conference presentations or manuscripts acknowledging students' contributions<br>• Student focus groups<br>• Teacher interviews |

## 4.4 Who Wants to Know?

Many people are currently invested in K-12 science education, including teachers, school administrators, students, their families, and their communities. Others, including practicing and retired scientists, state and federal agencies, and philanthropic organizations, have a stated interest in enhancing K-12 science learning. Broadly writ, these are the stakeholders in K-12 outreach and engagement. Yet, individual K12O&E efforts, especially those involving only a few individuals, cannot be all things to all people. The tough question becomes: who are the relevant stakeholders and what do they want to know? Guidelines for identifying stakeholders and their values and interests are available from a variety of sources (e.g., Felix et al. 2004; Frechtling 2002). If you have any interest in expanding your K-12 O&E activities, consider designing your evaluation with an eye to both current and future needs. Ideally, the evaluation plan has the flexibility and open-endedness to gather information that would be of interest to future stakeholders or larger or more stable funding sources.

## 4.5   What Resources are Available?

Although the term "resources" usually triggers thoughts of money, less obvious are institutional resources that already exist. Data from schools and districts, including student test scores or demographics and teacher retention rates or qualifications, may be available through web sites of school districts or state departments of education. Large school districts also may have evaluation personnel on staff. Universities may have personnel in similar evaluative or institutional research positions, or faculty with expertise in statistics, education research, or evaluation. Graduate students in these disciplines may also be looking for dissertation topics. Specific models for enhancing evaluation capacity have been built around tapping local resources. For example, the Howard Hughes Medical Institute's Peer Evaluation Cluster Initiative brings together science education program personnel to share their evaluation plans and activities and provide each other with critical yet constructive feedback (Felix et al. 2004, Hertle and Moreno 2004; York 2005). These small, more informal collaborations can serve as the foundation for larger efforts proposed in applications for extramural funding.

Evaluation can be done even in small scale efforts, like some of those described here. For example, a seventh grade teacher might collaborate with a scientist who is the mother of one of the children in his class. He knows that her expertise is in cardiac physiology, a topic in the middle school life science curriculum in his district. They want to start by figuring out what the students already know about the cardiovascular system, so at the start of the unit, they ask the students to share their thoughts about how and why blood circulates through the body. After brainstorming as a large group, the teacher and scientist ask the students to draw a picture of how oxygen gets to the different tissues throughout the body. After class, the teacher and scientist review the students' drawing, identifying common misconceptions and choosing lessons that they anticipate will help students develop a better understanding of the structure and function of the cardiovascular system. After the unit is completed, the teacher and scientist review the students' work, including their responses on an end-of-unit quiz, to determine where students were most likely to have misunderstandings. They use this information to revise the unit activities in preparation for next year. Although they did not have access to funding or special expertise, they are still able to evaluate their work in a way that is valuable for everyone involved. Other smaller scale approaches to collecting evaluation data include asking participants to write "minute papers," during which participants write their thoughts in response to a prompt question, or keeping a log of all email communications among partners (see Sundberg (2002) for additional examples).

## 4.6 How Will You Know It?

There is a range of methods commonly used for answering questions in the evaluation. Quantitative methods (which answer 'how many or how much change occurred') questions) include questionnaire surveys and checklists. Qualitative methods (which focus on "how?" "what?" and "why?" questions) include focus groups, semi-structured interviews and observation. Each method has strengths and weaknesses; the challenge is to identify the most appropriate method for the question (Designing an Evaluation, http://www.ngo-support.net/sw22918.asp. Accessed 2/2/2007).

The methods useful for gathering evaluation data depend on the evaluation's goals. If the aim is to develop an in-depth understanding of all of the ways a scientist models how to make inferences during a series of classroom visits, qualitative approaches like interviews with the teacher and scientist, focus group discussions with the students, and classroom observations by a third party observer may be the most useful strategies. In contrast, if the aim is to measure changes in students' interest in pursuing scientific careers after interning in laboratories for the summer, and to generalize the findings across a number of programs that involve student lab internships, a quantitative approach is likely to be more appropriate. "Mixed-methods" evaluations, which combine qualitative and quantitative approaches, can yield a more holistic picture and are more likely to capture unanticipated outcomes, for example:

The staff of the Partnership for Research and Education in Plants (PREP; www.prep.biotech.vt. edu) was implementing its first teacher-scientist meeting, which brought together all of the participating scientists and teachers. A series of small and large group discussions centered on how to better engage students in reflecting on their scientific findings. One scientist shared how she encouraged the graduate students and post-doctoral fellows in her group to reflect on their own and each other's findings by asking them each to give updates on their research during weekly group meetings. Although it was time consuming, she explained that the result made it worthwhile. Not only were her group members knowledgeable about each other's work, they more often offered each other constructive feedback. Before she started using this approach, she felt that she was the only one to offer such feedback. Upon hearing this, another scientist made the decision to reform the structure of her own group's weekly meetings. Excitingly, the PREP staff learned of this outcome because the meeting was being evaluated in an open-ended, qualitative way via the participants' written reflections, as well as comments made by participants during follow-up focus groups. Although this is a single example, this finding led to a more systematic evaluation of how participation in the partnership was influencing scientists' own practice, including the conduct of their research and the mentorship of their trainees.

Although a distinction is often drawn between formative and summative evaluation, the same methods and even the same data can be used for both. The difference is in

HOW the findings are used, to inform decisions (formative), for example how to revise an activity to address students' misconceptions, or to demonstrate impacts and outcomes of the program (summative), for example what new knowledge students have after completing a lesson.

Many different methods and tools can be used to collect and analyze data (Table 4.3). Teachers evaluate student learning in formal and informal ways on a daily basis and are likely to have good ideas or tools they have already developed to document student learning (e.g., assignments, tests, rubrics, etc.). Because there is not a common search engine or database for finding education evaluation and research literature, broad search engines like Google Scholar (http://www.scholar. google.com) can be useful. Specific sources for finding instruments include published literature and instrument databases such as:

- Mental Measurements Yearbook (Buros Institute; available through OVID, http://www.ovid.com/)
- Education Research Information Clearinghouse (ERIC; http://www.eric.ed.gov)
- Educational Testing Service's Test Link (http://www.ets.org/testcoll/)
- Educator's Reference Desk (http://www.eduref.org/)
- PsycINFO (http://www.apa.org/psycinfo/)
- Teacher Reference Center (TRC) through EBSCO Host (http://www.ebscohost. com/)

When choosing whether to use an existing instrument or develop a new one, consider whether the outcomes and impacts of your efforts are likely to be measurable using previously developed tools. Using an existing instrument can save time and resources, but the data must be examined to ensure any interpretations are valid (i.e., does the instrument measure what it was intended to measure?) and reliable (i.e., does the instrument perform predictably with similar populations such that the findings can be generalized?). Any time an instrument is changed (e.g., new items are added, the instrument is administered online vs. in hard copy, etc.) or used to collect data from a new population (e.g., the instrument was developed to assess undergraduate non-science majors' knowledge about evolution, but will be used to assess high school students' knowledge), any inferences made must be reexamined for validity and reliability, which often requires specialized expertise in statistics and education measurement. Sometimes, the point of evaluation is not to collect representative data that lead to generalized findings, so there may be little value in investing time and effort to do this. Rather, the priority may be on collecting evidence to make realistic recommendations for improvement of K-12 O&E activities.

## 4.7  Other Logistics of Evaluation

If evaluation data will be collected at schools, it is important to involve school district personnel as participants. Most districts have a point-of-contact for research, evaluation, and assessment. This individual can help administer instruments, serve

**Table 4.3** Evaluation methods, data, and analysis

| Methodology | Types of data | Examples of data analysis |
|---|---|---|
| Questionnaire | Likert scale measures of beliefs or attitudes (e.g., "Science is fun," where typically a score of "1" corresponds to strongly disagree, "5" is strongly agree, and "3" is neutral) | Factor analysis to confirm validity of the measures and identify underlying construct(s)<br>Calculations related to how well different questions on a test or survey perform (reliability statistics)<br>Simple paired comparison statistics to examine outcomes of partnership intervention (e.g., is there any different in student performance on pre/post-tests?)<br>Repeated measures analyses, if instruments are administered more than twice |
| Assessment | Instruments for assessing knowledge or ability using multiple choice, true/false, or short answer questions | |
| Interview | Transcripts of audio recordings made during interviews | Content or discourse analysis to identify ideas, points, or thoughts that are repeatedly noted during interviews or focus groups or in the text of written reflections<br>Development of a case study to examine and illustrate a complex scenario in depth (e.g., how do students interact with each other, their teacher, and a partner scientist during an inquiry lesson?) |
| Focus group | Transcripts of audio recordings made during focus groups; field notes taken during focus groups | |
| Written reflection | Participants' responses to prompt questions | |
| Document | Student work, workshop manuals, lesson plans, event agendas, email communications among partners | Content or discourse analysis to identify ideas, points, or thoughts that are repeatedly noted during interviews or focus groups or in the text of written reflections<br>Coding of responses according to a standard scale (what are students saying, writing, or doing and what does it imply?)<br>Scoring (can behaviors be quantified through the development and use of a defined rubric?)<br>Documentation of outputs (e.g., number of lessons developed, lessons used, email communications, etc.) |
| Observation | Video recordings of classroom or work shop activities; systematic observations of behaviors during teaching and learning | |

This table is not meant to be comprehensive, but rather to illustrate the variety of ways data useful in evaluating partnerships can be collected and analyzed. Additional detailed examples are available (e.g., Frechtling 2002, Sundberg 2002).

as advocate for the project within the district, and ensure compliance with district regulations. Because many evaluation activities occur in classrooms, teachers are often integrally involved in garnering the assent of participating students, consent of their families, and approval by school administration. They are also commonly responsible for administering instruments, which can become cumbersome for the already busy teacher. If teachers dedicate substantial non-work time to completing evaluation forms, it may be appropriate to offer compensation.

### 4.7.1  *External Evaluation*

Although a number of funders recommend or even require that evaluation be conducted by an external evaluator to ensure impartiality, steps can be taken to mitigate the bias that the individuals involved may bring to designing, conducting, and reporting the evaluation. Just as a science researcher administers an experimental intervention and interprets its outcomes, an evaluator or education researcher can administer an educational intervention and interpret its outcomes. The steps a scientist might take to reduce bias include controlling variables and conducting additional experiments to rule out alternative explanations. Conducting classroom-based studies involving random assignment to experimental versus control or comparison groups is challenging at best. Thus, social scientists, including education researchers and evaluators, have developed well-accepted strategies to reduce bias. For example, using multiple data sources and multiple data collection methods (i.e., triangulation), engaging multiple people with relevant but disparate perspectives in analyzing data, and discussing conclusions and evidence with study participants to ensure accuracy of the interpretation (i.e., member checking) can reduce bias and maximize trustworthiness and credibility of evaluation findings (see Anfara et al. 2002).

Some projects currently have or are seeking funds to hire an evaluator, which can help reduce bias and ensure access to the right experience and expertise. Options include:

- Employing an evaluator internally, which can be less costly and can enable the evaluator to become intimately familiar with the project.
- Establishing a contract with an external evaluator, which can help ensure objectivity in the evaluation and can provide flexibility to hire different individuals with expertise to accomplish particular tasks.

## 4.8  Research

New intellectual understandings can arise out of the very act of application (Boyer 1990).

Scientists who become regularly involved in K-12 outreach and education are often interested in publishing their work in scholarly venues, including journals of education research and practice. Ideally, such work is informed by and contributes to the larger field of teaching and learning research. However, the body of theoretical and practical research in education is sprawling, comprising more than 20,000 articles published each year in over 1,100 different journals (Mosteller et al. 2004). The absence of any unified, systematic mechanism for cataloging or accessing this information makes it nearly impossible for scientists to keep abreast of the literature on science teaching and learning, much less use it to inform their own work. When the challenge of locating articles of interest is coupled with the difficulty of comprehending the findings of an entirely different discipline, with epistemologies, cultures, and practices distinct from those of the science community (Feuer et al. 2002; Shavelson and Towne 2002), many scientists throw up their hands in frustration. There are strategies for developing working knowledge of the literature and resources that can be useful along the way (Dolan 2007).

A number of online databases provide access to education research citations. For example, Education Resources Information Center (ERIC; http://www.eric.ed.gov) is an internet-based digital library of education research and information that provides bibliographic records of journal articles and other education-related materials with sponsorship from the U.S. Department of Education. Print versions of ERIC information are published monthly in two formats: Resources in Education (RIE) and Current Index to Journals in Education (CIJE). Two other databases, Educational Research Abstracts Online (published by Routledge of Taylor & Francis Group; http://www.informaworld.com/smpp/title~content = t713417651) and Education Research Complete™ (published by EBSCO; http://www.epnet.com/thisTopic.php?marketID = 4&topicID= 639) provide abstracts for thousands of journals, books, and monographs, as well as full text for many journals and education-related conference papers. Finally, Google Scholar (http://scholar.google.com) enables searches of scholarly literature, including peer-reviewed papers, theses, books, pre-prints, abstracts, and technical reports. Google Scholar employs robotic spider software to crawl links to all scholarly articles publicly available on the World Wide Web. The company has standing agreements with academic publishers, professional societies, pre-print repositories, and universities, which have helped maximize the "findability" of relevant education scholarship (Dolan 2007).

Even if an article of interest is successfully located, it may not be decipherable by non-experts. As noted in Chap. 1, the field of science education is replete with discipline-specific terminology. In addition to the resources noted for understanding vocabulary most relevant to schools and classrooms, other resources useful for interpreting evaluative, applied, and theoretical research include *Fourth Generation Evaluation* (Guba and Lincoln 1989), *Qualitative Research & Evaluation Methods* (3rd edition, Patton 2002) and *Research Design: Qualitative, Quantitative, and Mixed Method Approaches* (2nd edition, Creswell 2003). Resources from the fields of psychology, sociology, and anthropology can also be useful. For example, the

*Online Dictionary of the Social Sciences* (http://bitbucket.icaap.org/; Drislane and Parkinson 2007), hosted by Canada's Athabasca University, has 1,000 entries covering sociology and related disciplines. *Explorations in Learning & Instruction: The Theory Into Practice Database* (http://tip.psychology.org/; Kearsley 1994–2007) includes descriptions of over 50 theories relevant to human learning and instruction. The *Web Center for Social Research Methods* (http://www.socialresearchmethods.net/), developed by William M.K. Trochim, a Professor in the Department of Policy Analysis and Management at Cornell University, is another clearinghouse of instructional and informational resources useful for teaching and learning research. These resources can help non-experts in better positioning themselves to locate, decipher, evaluate, and learn from the education research literature (Dolan 2007).

# References

Anfara Jr VA, Brown KM, Mangione TL (2002) Qualitative analysis on stage: Making the research process more public. Educ Res 31:28–38.

Creswell JW (2003) Research design: Qualitative, quantitative, and mixed method approaches (2nd ed). Sage Publications, Thousand Oaks CA.

Dolan EL (2007) Grappling with the literature of education research and practice. CBE Life Sci Educ 6:289–296.

Drislane R, Parkinson G (2007) Online dictionary of the social sciences. Athabasca University. http://bitbucket.icaap.org/. Accessed 28 August 2007.

Ercikan K, Roth WM (2006) What good is polarizing research into qualitative and quantitative? Educ Res 35:14–23.

Felix DA, Hertle MD, Conley JG, Washington LB, Bruns PJ (2004) Assessing precollege science education outreach initiatives: A funder's perspective. Cell Biol Educ 3:189–195.

Feuer M, Towne L, Shavelson R (2002) Scientific culture and educational research. Educ Res 31:4–14.

Frechtling J (2002) The 2002 user friendly handbook for project evaluation. National Science Foundation, Arlington VA. http://www.nsf.gov/pubs/2002/nsf02057/. Accessed 25 April 2007.

Guba EG, Lincoln YS (1989) Fourth generation evaluation. Sage Publications, Newbury Park CA.

Hertle MD, Moreno N (2004) Peer evaluation cluster: Perspectives of a program officer and a grantee participant. Ann Meet Natl Inst Health Natl Cent Res Resour Sci Educ Partnership Award, February 26–29, St. Louis, MO. http://www.nahsep.org/hertlesepa.ppt. Accessed 12 April 2007.

Joint Committee for Standards of Educational Evaluation (1994) The program evaluation standards: How to assess evaluations of educational programs. Sage Publications, Thousand Oaks CA.

Kearsley G (copyright 1994–2007). Explorations in learning & instruction: The theory into practice database. http://tip.psychology.org/. Accessed 16 June 2007.

Mosteller F, Nave B, Miech E (2004) Why we need a structured abstract in education research. Educ Res 33:29–34.

Patton MQ (2002) Qualitative research & evaluation methods (3rd ed). Sage Publications, Thousand Oaks CA.

Shavelson RJ, Towne L (Eds) (2002) Scientific research in education. National Academy Press, Washington DC.

Stufflebeam DL (2001) Evaluation models. New Dir Eval 89:7–92.

Sundberg MD (2002) Assessing student learning. Cell Biol Educ 1:11–15.

York P (2005) A funder's guide to evaluation: Leveraging evaluation to improve nonprofit effectiveness. Fieldstone Alliance, St. Paul MN.

# Chapter 5
# Sharing the News

## Disseminating Information about Your Work and its Impact

A number of venues provide opportunities to get feedback from other scientists and educators about your work, share the lessons you have learned, disseminate the materials you have developed, and contribute to the body of knowledge about science teaching and learning. Only by publicizing K-12 outreach and engagement work will interested individuals be able to avoid reinventing the wheel and lay a solid foundation for a unified expansion of the profession and discipline of K-12 outreach, engagement, and partnership.

## 5.1  What to Disseminate?

Do you want to recruit additional participants for your project? Do you and your colleagues want to offer your work as a model that could be replicated by other teachers and scientists? Do you want to share results revealed through project evaluation and research? Different aspects of your work will likely be of interest to different audiences, including teachers, scientists, education researchers, and potential funders. Determining which message you want to share and to what end will guide you in identifying the appropriate venues for doing so. If you would like to involve additional students, teachers, or scientists, consider how many individuals you would like to recruit as well as their geographic distribution. For example, a presentation at the annual meeting of the National Association of Biology Teachers is an ideal way to reach hundreds of interested life science teachers across the country. If you only have the capacity to work with a handful of local classrooms, then meeting with science coordinators in nearby districts may be more productive. Other networks for reaching K-12 teachers include Building a Presence for Science (http://science.nsta.org/bap/) and state chapters of the National Science Teachers Association (http://www.nsta.org/about/collaboration/chapters/default.aspx#chapterlist). These groups have points-of-contact and email, web-based, or hard copy newsletters through which K-12 O&E opportunities can be announced. Some will even share member mailing lists.

Likewise, many scientific societies provide mechanisms for communication among members who are interested in science teaching and learning. For example,

E.L. Dolan *Education Outreach and Public Engagement*,
DOI: 10.1007/978-0-387-77792-4_5 © Springer Science+Business Media, LLC 2008

the annual meeting of the American Association for the Advancement of Science (AAAS) dedicates one or more strands to K-12 and public science education. Other organizations having education working groups, like the Education Committee of the American Society of Plant Biologists. Other societies have committed significant financial and personnel resources to pre-college and public education by publishing a journal related to teaching and learning in the discipline, for example, *CBE – Life Sciences Education* of the American Society for Cell Biology.

Consider also the format within which you would like to share your work. Different stakeholders have varied priorities for K-12 education and are accustomed to gathering information in diverse ways. For example, state and local legislators may be most interested in reading a single-page "white paper" that includes personal and compelling examples of K-12 O&E outcomes. Other funders may be looking for quantitative data about student achievement, including statistical analyses and graphical displays of data. Individual teachers and scientists may be most interested in looking at student work directly to assess the outcomes of a specific lesson on students' learning so that they could design their next lesson. The format for disseminating your work should be designed with the intended audience and desired outcomes in mind.

## 5.2 How to Disseminate?

Most professional organizations have annual meetings, print or online peer-reviewed publications, and web sites that enable individuals to reach their membership (Table 5.1). For example, many teachers attend annual meetings of state or national science education organizations, which comprise "how-to" workshop sessions and keynote speakers on current issues in science and education. Many scientific societies have venues where members share their educational innovations and resulting impacts online, in journals, or at meetings. Some science associations may have dedicated K-12 education sessions at their annual meetings or subsidize the participation of K-12 teachers in meeting events. Education researchers and evaluators communicate through their own professional organizations, including general associations and those dedicated to science teaching and learning.

A number of professional organizations have an articulated mission to enhance K-12 science education and offer venues for sharing the outcomes of teaching practice (i.e., how-to), applied and theoretical research and essays (e.g., commentaries on current issues, public policy, etc.). While several scientific societies publish journals dedicated to education research and practice, others feature education-related work with their scientific publications. Also, all of these organizations host regular meetings that include education-related sessions and draw a regional, national, or international audience.

As in any field, education journals tend to specialize with regard to research questions and methodologies. Some journals feature descriptive essays, others theoretical research, some ethnographic studies, others statistical analyses, and so on.

**Table 5.1** Venues for dissemination regarding K-12 life science outreach and engagement

| Organization | Publication opportunities Title | Typical content |
|---|---|---|
| American Association for the Advancement of Science | Education Forum, *Science* | Applied, theoretical |
| American Institute of Biological Sciences | *BioScience* | Practice, applied, essays |
| American Physiological Society | *Advances in Physiology Education* | Practice, applied, essays |
| American Society for Biochemistry and Molecular Biology Education | *Biochemistry and Molecular Biology Education* | Practice, applied, theoretical |
| American Society for Cell Biology | *CBE – Life Sciences Education* | Applied, theoretical, essays |
| American Society of Microbiology | *Journal of Microbiology & Biology Education* | Applied, theoretical |
| American Society of Plant Biologists | Commentary, *The Plant Cell* | Practice, applied, essays |
| Association for Biology Laboratory Education | *Labstracts* | Practice, essays |
| Association for Science Teacher Education | *Journal of Elementary Science Education* *Journal of Science Teacher Education* | Applied, theoretical |
| Association of College & University Biology Educators | *Bioscene: Journal of College Biology Teaching* | Practice, applied, essays |
| Ecological Society of America | *ESA Bulletin* *Frontiers in Ecology and the Environment* | Practice, essays |
| Genetics Society of America | Genetics Education, *Genetics* | Practice, applied, essays |
| National Association for Research in Science Teaching | *Journal of Research in Science Teaching* | Applied, theoretical |
| National Association of Biology Teachers | *American Biology Teacher* | Practice, applied |
| National Science Teachers Association | *Journal of College Science Teaching* *Science and Children* *Science Scope* *The Science Teacher* | Practice, applied |

While several scientific societies publish journals dedicated to education research and practice, others feature education-related work with their scientific publications (journal titles are noted in italics while features within scientific publications are noted in plain text). Also, all of these organizations host regular meetings that include education-related sessions and draw a regional, national, or international audience.

Articles describing teaching strategies and curricular innovations are usually found in trade journals tailored to a specific teaching and learning audience (e.g., pre-college, undergraduate, graduate, science center/museum, etc.). Articles in which theories are tested or developed, including those intended to demonstrate relationships between instructional approaches and learning outcomes or to understand the cognitive, social, and cultural underpinnings of teaching and learning, are generally found in journals of applied and theoretical research. Regardless of the question of interest, journals can be identified that have a mission to feature this kind of work or to reach an audience with similar interests (Table 5.1; see Dolan (2007) for a more comprehensive list of venues for publishing about education research and practice).

In some venues, findings can be a catalyst for enhancing science learning. The K-12 education community values data that are relevant to decisions they must make regarding the adoption, improvement, and expansion of programs (Zmuda et al. 2004). Evaluations of K-12 O&E activities can contribute to this effort. In other venues, findings can add to the body of knowledge about K-12 O&E and science teaching and learning, while guiding practice in these fields. Local, regional, and national organizations host meetings where results can be shared (e.g., all of the organizations listed in Table 5.1; Dolan 2007). Regardless of the venue where results are shared, great care must be taken when discussing findings. Results that demonstrate correlations can be misinterpreted as evidence of causal relationships. In addition, the stakes are high because of the continued search for the elusive "silver bullet" that can be rapidly and widely adopted with little consideration of the students, teachers, schools, or communities involved (Carpenter 2000).

## 5.3   What Does the Future Hold?

Only by publicizing K-12 outreach and engagement work will we be able to learn from and build upon each other's efforts, entice others to get involved, and lay a solid and unified foundation for the development of a profession and discipline of K-12 outreach, engagement, and partnership.

> ...one big obstacle is that the scientific community, for all its handwringing about a scientifically illiterate public, still views outreach as a dubious activity for those on an academic career path (Mervis 2007).

This quote highlights the tension inherent in scientists' involvement in K-12 O&E. Yet, we are in the midst of a reinvention of the university. The accountability movement in K-12 schools is making its way into higher education, especially among public institutions like state universities (see The National Forum on College Level Learning [http://collegelevellearning.org/; Miller 2006). The public wants to know how their dollars are being spent. The rapid expansion of the federal debt will require reconsideration of spending at all levels, including the funding of life science research. At the same time, discoveries in the life sciences are revolutionizing the choices U.S. citizens have to make, from what foods they eat to what healthcare they utilize. Finally, advances in information technology are enhancing the accessibility of this information via numerous avenues, from Google searches to open courseware and distance learning programs (e.g., Massachusetts Institute of

Technology 2006). Public engagement will make transparent the goings-on of the scientific community and prepare our citizens to bring their own critical eye to evaluating the information that is available.

A supportive infrastructure is developing and a number of rewards are already available to those who heed Leshner's call for public "dialogue" (2007). Some institutions have reward systems in place to encourage and compensate faculty who dedicate time and energy to meaningful public engagement. Other institutions are dedicating significant resources to building the public engagement capacity of current and future science faculty. Many of these efforts have been initiated in response to challenges and expectations of extramural funding agencies. As a result, paths are being forged that blur the line between the scientific and education communities. For example, NSF's Discovery Corps Fellowships (DCF; http://www.nsf.gov/funding/pgm_summ.jsp?pims_id = 6676) support post-doctoral work across disciplinary boundaries and must significant research and service components. Fellows, including those who have just completed their doctoral work and mid-career scientists who are considered senior fellows, can choose to affiliate themselves with existing NSF-funded K-12 O&E programs, such as the Graduate Teaching Fellows in K-12 Education (GK-12) program. The DCF request for proposals (http://www.nsf.gov/pubs/2007/nsf07516/nsf07516.htm) provides examples of service components directly related to K-12 and public science education, for example, partnering with an NSF Center for Learning and Teaching to "enhance the quality of secondary school chemistry laboratories" or with "artists or historians to develop a novel interdisciplinary collaboration that enhances public science education and highlights the connections between science and society."

As the rewards multiple and the demand for tools increases, the time is ripe for the establishment of the profession of K-12 outreach, engagement, and partnership (Dolan and Tanner 2005; Fraknoi 2005; Tanner et al. 2003). The science and education communities as a whole need to rethink their infrastructure to support this burgeoning field, from revisiting promotion and tenure guidelines to reinventing hiring practices (Bush et al. 2006; Lally et al. 2007; Talanquer et al. 2003; Tanner and Allen 2006; Tomanek et al. 2003). The groundwork is being laid for problem solving and knowledge sharing across the K-20+ continuum of science learning.

# References

Bush SD, Pelaez NJ, Rudd JA, Stevens MT, Williams KS, Allen DE, Tanner KD (2006) Approaches to biology teaching and learning: On hiring science faculty with education specialties for your science (not education) department. CBE Life Sci Educ 5:297–305.

Carpenter WA (2000) Ten years of silver bullets: Dissenting thoughts on education reform. Phi Delta Kappan 81:383–389.

Dolan EL (2007) Grappling with the literature of education research and practice. CBE Life Sci Educ 6:289–296.

Dolan EL, Tanner KD (2005) Moving from outreach to partnership: Striving for articulation and reform across the K-20+ science education continuum. Cell Biol Educ 4:35–37.

Fraknoi A (2005) Steps and missteps toward an emerging profession. Mercury 34:19–25.

Lally D, Brooks E, Tax FE, Dolan EL (2007) Sowing the seeds of dialogue: Public engagement through plant science. Plant Cell 19:2311–2319.

Leshner AI (2007) Outreach training needed. Science 315:161.

Massachusetts Institute of Technology (2006) OpenCourseWare marks 5 years. TechTalk 50 (22). http://web.mit.edu/newsoffice/2006/ocw.html. Accessed 28 November 2007.

Mervis J (2007) Public science: Pilot NSF program flies into stiff community headwinds. Science 318:1365–1367.

MA Miller (2006) Assessing college-level learning. Natl Cen Public Pol Higher Educ Policy Alert, May issue. http://www.highereducation.org/reports/pa_aclearning/nam.shtml. Accessed 26 November 2007.

Talanquer V, Novodvorsky I, Slater TF, Tomanek D (2003) A stronger role for science departments in the preparation of future chemistry teachers. J Chem Educ 80:1168–1171.

Tanner KD, Allen D (2006) Approaches to biology teaching and learning: On integrating pedagogical training into the graduate experiences of future science faculty. CBE Life Sci Educ 5:1–6.

Tanner KD, Chatman L, Allen D (2003) Approaches to biology teaching and learning: Science teaching and learning across the school–university divide: Cultivating conversations through scientist–teacher partnerships. Cell Biol Educ 2:195–201.

Tomanek D, Talanquer V, Novodvorksy I, Slater TF (2003) Responding to the call for change: The new college of science teacher preparation program at the University of Arizona. Cell Biol Educ 2:29–34.

Zmuda A, Kuklis R, Kline E (2004) Transforming schools: Creating a culture of continuous improvement. Assoc Superv Curric Develop, Alexandria VA.

# Index

Printed in the United States
By Bookmasters